Society for Arabian Studies Monographs No. 11

Series editors: D. Kennet & St J. Simpson

Water Management: The Use of Stars in Oman

Harriet Nash

BAR International Series 2237
2011

Published in 2016 by
BAR Publishing, Oxford

BAR International Series 2237

Society for Arabian Studies Monographs No. 11
Series Editors: D. Kennet & St J. Simpson

Water Management: The Use of Stars in Oman

ISBN 978 1 4073 0799 2

BAR Publishing is the trading name of British Archaeological Reports (Oxford) Ltd. British Archaeological Reports was first incorporated in 1974 to publish the BAR Series, International and British. In 1992 Hadrian Books Ltd became part of the BAR group. This volume was originally published by Archaeopress in conjunction with British Archaeological Reports (Oxford) Ltd / Hadrian Books Ltd, the Series principal publisher, in 2011. This present volume is published by BAR Publishing, 2016.

Printed in England

BAR
PUBLISHING

BAR titles are available from:

BAR Publishing
122 Banbury Rd, Oxford, OX2 7BP, UK
EMAIL info@barpublishing.com
PHONE +44 (0)1865 310431
FAX +44 (0)1865 316916
www.barpublishing.com

Society for Arabian Studies Monograph Series

Series editors: D. Kennet & St J. Simpson

The *Society for Arabian Studies Monograph Series* was launched in 2004 with the intention of encouraging the publication of peer-reviewed monographs on the archaeology, early history, ethnography, epigraphy and numismatics of the Arabian Peninsula and related matters. Creating a specific monograph series within the *British Archaeological Reports International Series* is intended to allow libraries, institutions and individuals to keep abreast of work that is specifically related to their areas of research. Whilst research and conference volumes in the series will all be peer-reviewed according to normal academic procedures, the decision was taken to allow the publication of doctoral theses, field reports, catalogues and other data-rich work without peer review where this will permit the publication of information that, for one reason or another, might not otherwise be made available.

Already published:

BAR –S1248, 2004 Sasanian and Islamic Pottery from Ras al-Khaimah *Classification, chronology and analysis of trade in the Western Indian Ocean* by Derek Kennet with a contribution by Regina Krahl. Society for Arabian Studies Monographs No. 1. ISBN 1 84171 608 1.

BAR –S1269, 2004 Trade and Travel in the Red Sea Region *Proceedings of Red Sea Project I held in the British Museum October 2002* edited by Paul Lunde and Alexandra Porter. Society for Arabian Studies Monographs No. 2. ISBN 1 84171 622 7.

BAR –S1395, 2005 People of the Red Sea *Proceedings of Red Sea Project II held in the British Museum October 2004* edited by Janet C.M. Starkey. Society for Arabian Studies Monographs No. 3. ISBN 1 84171 833 5.

BAR –S1456, 2005 The Tihamah Coastal Plain of South-West Arabia in its Regional Context c. 6000 BC – AD 600 by Nadia Durrani. Society for Arabian Studies Monographs No. 4. ISBN 1 84171 894 7.

BAR –S1661, 2007 Natural Resources and Cultural Connections of the Red Sea *Proceedings of Red Sea Project III held in the British Museum October 2006* edited by Janet Starkey, Paul Starkey and Tony Wilkinson. Society for Arabian Studies Monographs No. 5. ISBN 9781407300979

BAR –S1776, 2008 La Péninsule d'Oman de la fin de l'Age du Fer au début de la période sassanide (250 av. – 350 ap. JC) by Michel Mouton. Society for Arabian Studies Monographs No. 6. ISBN 978 1 4073 0264 5

BAR –S1826, 2008 Intercultural Relations between South and Southwest Asia *Studies in commemoration of E.C.L. During Caspers (1934-1996)* edited by Eric Olijdam and Richard H. Spoor. Society for Arabian Studies Monograph No. 7. ISBN 978 1 4073 0312 3

BAR –S2052, 2009 Connected Hinterlands *Proceedings of Red Sea Project IV held at the University of Southampton September 2008* edited by Lucy Blue, John Cooper, Ross Thomas and Julian Whitewright. Society for Arabian Studies Monographs No. 8. ISBN 978 1 4073 0631

BAR –S2102, 2010 *Ports and Political Power in the Periplus Complex societies and maritime trade on the Indian Ocean in the first century AD* by Eivind Heldaas Seland. Society for Arabian Studies Monograph No 9. ISBN 9781407305783

BAR –S2107, 2010 Death and Burial in Arabia and Beyond *Multidisciplinary perspectives* edited by Lloyd Weeks. Society for Arabian Studies Monograph No. 10. ISBN 9781407306483

Potential contributors

Please contact the editors in the first instance:

Dr Derek Kennet: Department of Archaeology, Durham University, South Road, Durham, England DH1 3LE
Derek.Kennet@durham.ac.uk

Dr St John Simpson: Department of the Middle East, The British Museum, London, England WC1B 3DG
ssimpson@thebritishmuseum.ac.uk

Acknowledgements

I am indebted to many organisations and people for their support and advice. I especially thank my supervisor, Dionisius Agius, who encouraged me to undertake this research, and has ever since been a source of support and inspiration. His wife, Anne, has also been most generous throughout the study, and my family helped in many ways, especially Marius Albu, Sue Nash and Ian Warwick.

The many star users, *falaj* officials and others interviewed in the study areas in Oman are mentioned in the relevant sections. Without them this work would not have been possible and I am deeply grateful for their assistance and patience.

Ahmed Al-Mukhaini, Technical Secretary of the Historical Association of Oman (HAO) provided much useful advice and many contacts at the start of the project. I took advice on matters of Islamic jurisprudence from Abdul Rahman Al-Salimi, Editor of the journal Al-Tasamoh in the Ministry of Awqaf (mosque properties) and Religious Affairs, and on astronomy from Marwan Shwaiki, Director of the PDO (Petroleum Development of Oman) Planetarium. David Whitfield provided sustenance in abundance, introduced me to several people who assisted with the study, and was with me on the fateful day in Mudaybi and Zahib when I first saw a building used for stargazing. Hilal Al-Siyabi (HAO) arranged for several Arabic manuscripts to be typed into Microsoft Word. Abdullah Al-Jafri, manager of the government pottery in Bahla, spent much time patiently explaining terms used in *falaj* management and date cultivation. Ali Said Al-Yahyai, a retired *falaj* digger, and Abdullah Al-Jafri also helped me with contacts in the *falaj* organisation in Bahla, and I thank both them and their families for befriending and taking care of me.

This study would not have been possible without the unstinting support of the Ministry of Water Resources (MWR), which provided accommodation in Al Hamra and Ibra and assistance from its regional offices, especially for introductions to *falaj* officials. I am particularly indebted to Hamad Al-Hatmi, Director of Aflāj and Support Wells, Saif Al-Shaqsi, Advisor on Water Resources to the Minister and Zaher Al-Suleimani, Director General of Water Resource Assessment, for their advice and support. I would also like to thank Suleiman Al-Abri, Director of Water Resources in Nizwa and Mohammed Al-Adawi, Assistant Director; Ahmed Al-Sabbari, Head of Water Resources in Al Hamra, Hamid Al-Sabbari, Head of Water Resources in Bahla; Mohammed Al-Hashami, technician in Izki; Ibrahim Al-Namani and Ali Al-Habsi, technicians in Mudaybi; Rashid Al-Tawqi, Head of Water Resources in Mudaybi. Saleh Al-Adhary in the planning department in Muscat spent many hours preparing maps for me to use, and Tariq Helmi helped with illustrations and the provision of statistical data. Saif Al-Amri, Deputy Director, Aflaj and Support Wells Department and many others in the Ministry offices in Muscat helped by providing information and data. Yusuf Al-Kindi, Director of Al Hamra's Municipality, kindly assisted with accommodation in Al Hamra, and I am also grateful to the watchmen at the Municipality, and to Abdullah Khalili, driver, for numerous cups of tea and coffee.

At the Department of Agriculture at Sultan Qaboos University (SQU), Professor Prathapur arranged for me to be treated as a visiting scientist, allowing me to use the University facilities. Abdullah Al-Ghafri, an expert on Oman's *aflāj*, has helped in many ways, stimulating ideas, giving me access to his extensive library, use of his office and advice and encouragement throughout the study.

Several people helped me in the Ministry of Heritage and Culture, including Biuba bint Ali Al-Sabiri, Director of Excavations and Archaeological Research, Khair bin Antar Al-Busaidi, General Manager, Department of Manuscripts and Documents, and Rahma Al-Farsi, Head of Archaeology. Particular thanks go to Mohammed Al-Tarshi, Head of Manuscripts, for his assistance in locating and scanning documents, and to Ahmed Al-Tamimi, in charge of the restoration of Bahla Fort, who helped with information on manuscripts related to *aflāj* and showed me the graffiti in Bahla mosque.

Walid Al-Tikriti, Department of Archaeology and Tourism, Al Ayn (UAE), spent several days showing me the archaeological sites in the area of Al Ayn, in particular the 1st millennium BC *aflāj* he had excavated, giving me much needed insight into the development of these systems in eastern Arabia.

I am also grateful to Suleiman Al-Riyami of Red Wave Trading computer services in Al Hamra and Bahla for his help in times of trouble; to Mattab bin Sultan Al-Habsi for accommodation when the Rest House in Sinaw was closed for restoration; to Al Maha car hire, Al Sheiba Opticals in Bahla, the Tourist Rest House in Sinaw, the Corniche and Marina Hotels in Muscat for their excellent service.

This research would not have been thought of if I had not worked on the Al-Hajar project in March 2005. For this I am indebted to Jocelyn and Jeffery Orchard. Several people mentioned above assisted me at that time; others include Ahmed Al-Shukaili, who introduced me to *falaj* officials in Bisya and to the library at Sultan Qaboos University; Yaqub Al-Kamiali and Nasr Al-Salmi in the *falaj* section of the MWR in Nizwa; Said Al-Harrasi, Head of Special Collections, Sultan Qaboos University Library; Ali Al-Mahroqi (HAO and Ministry of Heritage and Culture) who, with his uncle Mohammed, showed me how the sundials are used and *falaj* irrigation carried out in Adam.

I did not meet the astronomers of the Royal Court until near the end of my fieldwork in 2008, but am, nevertheless, grateful to Mohammed Salim Al-Busaidi and Suleiman Hallal Al-Busaidi for their enthusiasm for the project and the loan of star identification equipment.

John Wilkinson helped with suggestions for the scope of works at the beginning of the project and permission to use his archival material lodged in the library of the University of Exeter; Daniel Varisco and Paul Kunitzsch: with advice on star names; Paul Auchterlonie in tracking down literature; Qaisar Iskander helped with fonts; Mike Edmunds provided numerous references for funding, sometimes at very short notice.

The Seven Pillars of Wisdom Trust provided a grant of £2000, which covered my expenses for fieldwork in Oman in January-February 2006 and the cost of some equipment. For the second and third years of the study I received a maintenance grant from the Arts and Humanities Research Council, together with a contribution to the expenses of fieldwork. A Prince Al-Waleed Furthering Understanding grant of £1000 covered part of my expenses for a visit to Oman in November 2008 to consolidate links with academics in Sultan Qaboos and Nizwa Universities. An exhibition including the study results held at Bait Al Baranda museum in Muscat could not have taken place without the commitment of the museum's director, Malik Al-Hinai and his team. Steve McNulty of the British Council introduced me to Malik, and provided a PowerPoint projector both for the exhibition and subsequent presentations in Oman.

I am also greatly indebted to Roger Clark of CL Voelcker, where I currently (2011) work as a hydrogeologist, for supporting the study and to other colleagues, namely Keith O'Brien and Janice Sheldon for assistance with illustrations, and André Smit for advice on night photography.

System of Transliteration of Arabic Words

Arabic words are generally romanised using the Library of Congress system below, with a few extra vowels to more closely represent the spoken language given in square brackets.

Consonants

’	ء	ṭ	ط
b	ب	ẓ	ظ
t	ت	ʿ	ع
th	ث	gh	غ
j [g,y]	ج	f	ف
ḥ	ح	q	ق
kh	خ	k	ك
d	د	l	ل
dh	ذ	m	م
r	ر	n	ن
z	ز	h	ه
s	س	w	و
sh	ش	y	ي
ṣ	ص		
ḍ	ض		

Vowels

Long	
ā [ē]	ا
ū	و
ī	ي
Doubled	
uww (final ū)	وّ
iyy (final ī)	يّ
Short	
a [e])	َ
u, [o]	ُ
i, [e]	ِ
a (t in construct)	ة
Dipthongs	
ay	يْ
aw, (ō)	وْ

Exceptions are conventional spellings of place names (eg towns, villages and *aflāj*) and personal names. *Nisba* endings (relative adjectives describing a person as "of" a particular place or attribute) for proper nouns are, therefore, given as /i/ not /ī/, and the letter *ʿayn* and the definite article */al-/* are often dropped at the beginning of a word. Some little used Arabic plurals, such as *sihām* (pl. of *sahm*) and *āthār* (pl. of *athar*), both water/time divisions, are given as the singular with an "s" after, as in *athar*s.

Where local pronunciation differs from Modern Standard Arabic (MSA), the sound is given in the closest possible English equivalent. For example, the letter *alif* is transcribed as /ē/ for the spoken long vowel between those in "fast" and "bled", as in Yēh for the star Polaris (جاه).The hard *jīm*, transcribed as /g/, is widely used in Oman except for certain words, such as *najm* (star) and in many star names; sometimes it is pronounced as /y/ (as in *nayūm* for *nujūm* – stars). However many words spoken with a hard *jīm*, such as *falaj* (the irrigation system that is the subject of this work) are commonly transcribed with /j/ and not /g/, and this custom is followed here, which means that the romanised form is not necessarily a guide to pronunciation. The sounds /dh/, /ḍ/ and /ẓ/cannot always be distinguished, and at the ends of words, *tā' marbūṭa*, *alif*, *alif maqṣūra* and *hamza* preceded by *alif* or *fatḥa* are often indistinguishable in speech, which has led to a variety of written forms.

For words starting with letters that replace the /l/ of the article, the transcription of speech generally gives the sound as heard, as in ath-Thurayyā. However, the Modern Standard Arabic article /ال / is expressed dialectically in many ways, such as /al-/, /el-/, /la/ or /le/. The article of star names is, therefore, often omitted in order to make the text and tables easier to read, and avoid the necessity to always distinguish between written (rare) and spoken forms. Hence star names with adjectives sometimes appear incorrectly as a possessive construct (*iḍāfa*), but the true meaning of star and adjective is clear. For citations and bibliographical references, the names of Arab authors writing in English are given as commonly seen in print, with the article /Al-/ preceding the family/tribal name, and no diacriticals, as in Al-Ghafri. By contrast, for authors of mediaeval and modern Arabic texts, the article /al-/ and full diacriticals are given, as in al-Kindī.

Transliterated or transcribed words are given in italics, with the exception of proper nouns, including star names, and words that have entered the English language, such as wadi (an ephemeral river or stream and its valley). Although the term *falaj* has appeared in Chambers Dictionary since 1988, it is not in common use and has therefore been italicised.

The presentation of Omani personal names varies. The formal form with several generations each preceded by *bin* (son of) is not always used: among professionals it is now common to drop the *bin*s and have only the first name and family/tribal group name, as in Hamad Al-Hatmi. In less formal surroundings, as in study villages where there is only one tribal group, people are often called by their first name and that of their father, again with no *bin*, as in Saif Salim.

Abbreviations

Abbreviations in common use, such as for dating systems, are given in full here but not in the text. Others, such as Ministry names, are given in full on the first use in each chapter and listed here in addition for ease of reference.

AD	*Anno Domini*
AH	*Anno Hijrae*
ASL	above (mean) sea level
BC	before Christ
BP	before present
C	century, centuries
CA	Classical Arabic
GPS	global positioning satellite equipment
HAO	Historical Association of Oman
K	Kelvin: scale of absolute temperature (0 Kelvin = -273.16° Celsius)
MHC	Ministry of Heritage and Culture, Oman
MRW	Ministry of Water Resources, Oman
MRMWER	Ministry of Regional Municipalities, Environment and Water Resources, Oman
MRM&WR	Ministry of Regional Municipalities and Water Resources, Oman
MSA	Modern Standard Arabic
SQU	Sultan Qaboos University, Oman
UAE	United Arab Emirates
N	north
S	south
E	east
W	west
b.	*bin* (son of); for authors of Arabic texts and historical figures
c.	*circa* (approximately)
cont$^{d.}$	continued
d	day
d.	died
ha	hectare
hr	hour
m	metre/month
min	minute
n.k.	not known
pl.	plural
s.	singular

The use of the abbreviations MRW, MRMEWR and MRM&WR needs further explanation:

- In 1999 the Ministry of Water Resources (MWR) took over responsibility for water resource assessment, improvement and regulation;
- In 2001, the MWR became the water resources section of the Ministry of Regional Municipalities, Environment and Water Resources (MRMEWR);
- In 2007, the environment section was established as a separate ministry, leaving the water resources section in the Ministry of Regional Municipalities and Water Resources (MRM&WR).

Whichever the organisation, it is the water resources section that I have been dealing with during this research. The title Ministry of Water Resources and the abbreviation MWR are therefore used to refer to all three organisations. The official titles are, however, retained for reference to specific publications.

For dates related to activities or literature in the Muslim world, Hijrī dates precede dates in the Christian calendar, for example $4^{th}/10^{th}$ C means 4th century AH and 10th century AD. Single dates refer to the Christian calendar, eg 20th C means the 20th century AD.

TABLE OF CONTENTS

APPENDICES

LIST OF TABLES

LIST OF FIGURES

All photographs are by the author.

ABSTRACT

Ancient settlements in Oman are frequently found where crops can be cultivated with groundwater distributed to fields by gravity flow. These irrigation systems are called *aflāj* (s. *falaj*), and they are still pivotal to community life: in 1997 over 3000 such systems were still in use. The allocation of *falaj* water among farmers traditionally depended on the use of the sun and the stars to track the passage of time. This practice, and particularly the use of stars, is disappearing rapidly: the availability of watches since c. 1970 means that there is now little technical reason to use the stars.

The study reported in this monograph is, with minor changes, the author's PhD thesis, the degree being awarded in March 2009. It presents fast disappearing information on the ethnography of agricultural communities in northern Oman and especially on the use of stars. It aims to identify the stars used and record the methods of stargazing in sufficient detail that the systems could be replicated if all local knowledge were lost. It also aims to raise awareness and improve the understanding of the significance of the traditional use of the stars in irrigated agriculture. This aspect of the cultural heritage of the region has not been written about previously in such detail.

The research questions concern how, where and why stars are still used.

There are several documents on *falaj* building and organisation in Oman, but relatively little on telling the time, and much less on the stars than on the use of the sun. The available documents were reviewed, but the main source for this study is data collected in the field, interviewing *falaj* managers and stargazers and watching the stars with them. Fieldwork was carried out in different seasons since the stars used vary through the year. International experts were consulted on various aspects of the work, in particular Daniel Martin Varisco on folk astronomy and Paul Kunitzsch on star names.

The study focused on five settlements where stars are used to this day: Qarya Beni Subh (Qarya), close to the mountains near the town of Al Hamra in the Interior Region; Al Fath, Zahib, Sudayra and Barzaman near the towns of Mudaybi and Sinaw in the Eastern Region. Additional information was obtained from a few other places for comparison, and during the course of the research it was found that only about eight of the 3000 active *falaj* communities still use stars for time keeping.

The methods of stargazing are described in detail, together with the names of the stars in Oman and their international classification where possible. It was found that the number of stars, the stars used, and their names vary from place to place. The time that each star represents is more variable than the difference in length of night during the year. The stars used for *falaj* purposes are compared with those used in the region for navigation and with the stations of the moon of the Islamic calendar, but no evidence of any link between them was identified. On the basis of the history of *falaj* systems in Oman, the stars used and the star names, several of which are from the pre-Islamic Bedu tradition, it is concluded that the system of star use for timing water dates from pre-Islamic times, before the introduction of the stations of the moon.

The main reason for the continued use of stars by a few communities, even though wristwatches are now widely available, is thought to be adherence to tradition both for its own sake and to avoid disputes over the sharing of water. It is considered unlikely that this ancient practice will survive more than ten or twenty years unless the younger generation takes an interest in learning, and putting into practice, the traditions of their forebears.

1 INTRODUCTION

1.1 ORGANISATION OF THE MONOGRAPH

The monograph is presented in four parts:
i. Chapters 1 to 4: introductory;
ii. Chapters 5 to 10: presentation and analysis of information;
iii. Chapters 11 and 12: discussion and conclusions;
iv. Reference information including appendices, glossary, bibliography and index of star names.

The introductory section starts in this chapter with an outline of the objectives and significance of the research, a description of its scope and the programme of the research including the selection of study areas, methods of collecting and interpreting information, tools used and constraints on the type and quality of information collected. Documentary sources of information including the available literature are reviewed in terms of their significance to the study in Chapter 2. The basics of astronomy and its history are given in Chapter 3, and in Chapter 4 the irrigation systems that are the subject of this study are described in general terms. Chapters 3 and 4 provide an essential basis for understanding and interpreting the information collected during this study.

Chapters 5 to 10 present the detailed information collected in the study areas. Chapter 11 draws together the assessments made in individual chapters, and expands on them, relating them to each other and information from elsewhere, while Chapter 12 relates the findings to the conceptual framework and the research questions and objectives.

1.2 GEOGRAPHICAL SETTING

The Sultanate of Oman occupies an area of 309,500 square km (Ministry of Information, 2007) in the east of the Arabian Peninsula. It straddles the Tropic of Cancer and is bordered to the west by the United Arab Emirates and the Kingdom of Saudi Arabia, to the south by the Yemen Arab Republic, and to the east by the Gulf of Oman and the Indian Ocean. The location is shown in Figure 1.1.

The main physical regions are:
- mountain ranges: the Hajar Mountains, which run parallel to the coast in the north of the country, and the Dhofar Mountains in the south, which run parallel to the coast inland of the Salalah plain;
- foothills of the mountains and gravel outwash plains: mainly in the north of the country, south of the Hajar Mountains;
- stony desert plains (the Nejd), occupying much of the interior;
- sand seas: in the west, the eastern part of the Rub al-Khali, and in the north west, the Wahiba Sands;
- coastal plains of which the Batinah in the north is the most extensive.

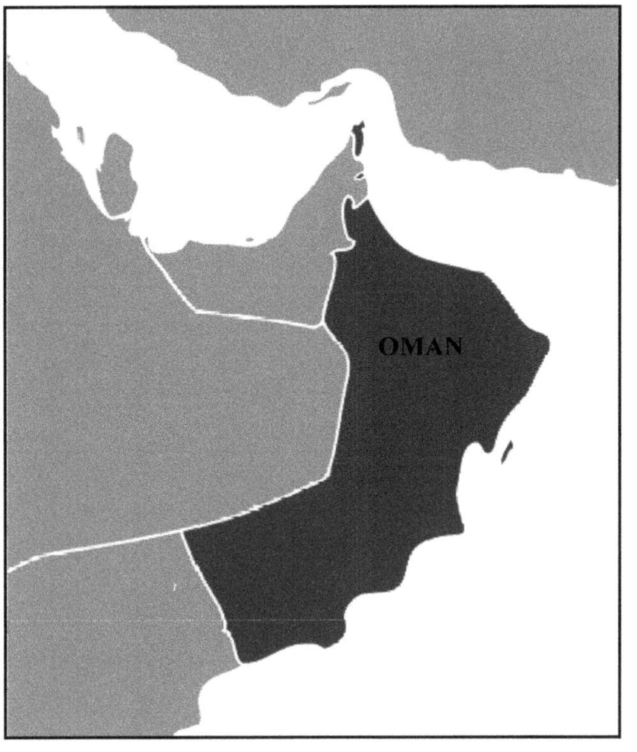

Figure 1.1 Location of Oman
based on map from Wikimedia Commons © Goldsztajn
The international boundaries shown are schematic

The annual average rainfall is 100 mm/yr or less over most of the country, rising to 350 mm/yr in the mountains (MRMEWR, 2005: 26). In the south of the country most precipitation occurs during the summer monsoon, between June and September, while in the mountains in the north rain may fall at any time of year. Potential evapotranspiration is much greater than the rainfall, reaching approximately 3 m per annum.

From the 3rd millennium BC, Oman has been an important centre for maritime trade (eg Agius, 2008) and is known for her export of copper from the earliest times. She has, therefore, had access to the culture and technology of all of the great Old World civilisations throughout her history. Oil was discovered in the early 1950s and was first exported in the second half of the 1960s. Before His Majesty Sultan Qaboos succeeded his father in 1970, the country had only three non-Qur'ānic schools, ten kilometres of tarmac road and a population of 658,000 (Ministry of Information, 2007). From 1970 development was rapid, with the establishment of government schools and health care for all.

1.3 OBJECTIVES AND SIGNIFICANCE OF THE RESEARCH

The timing of *falaj* water shares among farmers once depended almost everywhere on the use of the sun and the stars. This practice, and particularly the use of stars, is fast disappearing, together with the knowledge about the stars and how they are used: of approximately fifteen *falaj* systems visited in March 2005, when the author was when taking part in the Al-Hajar archaeological project[1], stars were still being used in only one – Zahib, near Mudaybi. It is therefore important to the preservation and knowledge of the cultural heritage of Oman, and the Arabian Peninsula, to record where, how and why stars are still used and to develop an understanding of this ancient practice.

The purpose of the research was to record and assess the practice of using stars to time the allocation of irrigation water in Oman before it dies out, and in sufficient detail to be able to replicate the system of stargazing should it disappear in the future. A further aim was to increase awareness, both in Oman and internationally, about this important cultural aspect of water management. The following questions were addressed:

How do the stars help farmers to distribute water?
Knowledge of the methods used to track time by the stars, to determine when individuals are allowed to use their water rights at night, may soon become extinct. Recording which stars people watch, how they watch them and how the information is transferred to people using the water was fundamental to the research.

Where and why are stars still used to track the passage of time for the allocation of irrigation water?
The ready availability of watches mean that there is now little technical reason to use the stars. The fact that they are still used in a few places is probably due to attachment to tradition. Adherence to tradition is one way to minimise disputes, which could disrupt the peaceful, productive life of the *falaj* community. This leads to the sub-question:
How are disputes over the allocation of water resolved?
One method of avoiding disputes is to record people's water rights in writing, so a related sub-question is:
What written records are available regarding the allocation of water?
The main components of the conceptual framework are illustrated in Figure 1.2. Productivity and security are achieved by the efficient utilisation of a scarce resource.

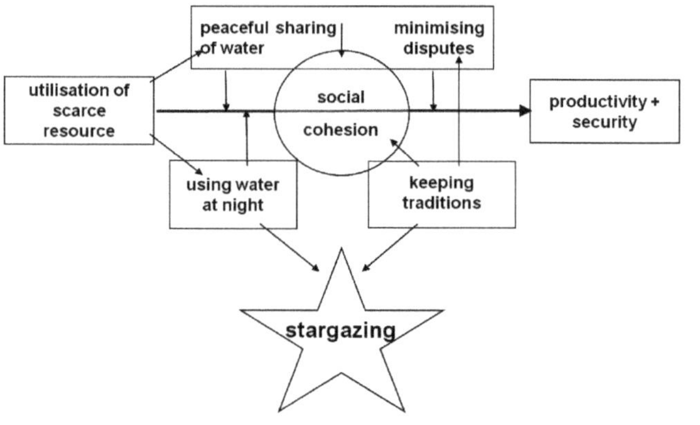

Figure 1.2 Conceptual framework

Falaj water, unlike water from wells, flows continuously, and irrigation is therefore carried out 24 hours a day. The need to time the distribution is considered in Section 4.5. Currently stars are not the only means to time the water distribution: wristwatches are now widely used. Hence stargazing *per se* is no longer needed to achieve the aim of agricultural production, but keeping to traditional practices is clearly important in some communities.

Although the overall aims of *falaj* management can be achieved without the traditional methods of time keeping, it is still necessary for the distribution of water to be agreed by all, and for all *falaj* users to co-operate in running the system. A number of factors may be involved in the adherence to tradition, including:

- fear of change
- fear of disputes over the allocation of water
- potential loss of the sense of identity and security.

The allocation of *falaj* water is sometimes disputed among farmers. The area of water rights in Oman and resolving disputes has been written about since the late 3rd/9th - early 4th/10th C (al-Ḥawārī, 1975). Currently, it appears that records are kept by communities mainly where disputes may arise.

This is the first time that many of the stars used for the *aflāj* in Oman have been identified by their international names, or the methods used for telling the time by stars described in detail. It therefore represents a significant contribution to the knowledge of, and provides a unique resource for people interested in, the cultural history and folk astronomy of the Arabian Peninsula.

[1] An archaeological project directed by Jocelyn Orchard, Jeffery Orchard and Jennifer Scarce operating in northern Oman since 1980.

1.4 STUDY AREAS AND THEIR SELECTION

The villages selected for the study are in two separate areas, shown in Figure 1.3. The first is the area of Al Hamra, a town close to the mountains in the Dakhiliyah (Interior) Region. The second is about 80 km from the mountains near the town of Mudaybi in the Sharqiyah (Eastern) Region. Zahib, near Mudaybi, was identified for inclusion in March 2005. Given the large numbers of *falaj* systems still in use in Oman (Chapter 4), it was considered impracticable to identify every settlement still using stars. In addition, the programme for the PhD did not allow the time to study more than a few villages at the level of detail needed to meet the aims of the research. It was therefore decided to focus on:

- the area south of the Hajar Mountains, which is less affected by modern developments, such as irrigation by pumped wells, than to the north, and
- *falaj* systems that have not already been studied in any detail, even if not for the method of star use.

The search for other *falaj* systems in addition to Zahib to include in the study started in Bahla, where there was a support base of people willing to help me following my stay with the Al-Hajar Archaeological Project. Qarya Beni Subh (Qarya) was the only village found in the areas of Bahla and Al Hamra, as defined by the responsibilities of the Ministry of Water Resources (MWR), to still use stars for the *falaj* water distribution. I therefore decided to collect information about the former use of stars in a few nearby settlements, Al Hamra, Qala and Misfa for background.

Several people consulted in the initial stages of the search told me of places still using the stars. Many of these were checked without success, including Birkat al-Mawz near Izki, Mintirib in Bediya District, Habbi near Bahla and Sinaw near Mudaybi. However, Ibrahim Al-Namani, a *falaj* technician in the Ministry of Water Resources (MWR) in Mudaybi knew of several villages in the area using stars, in addition to Zahib, and identified three others: Al Fath, Sudayra and Barzaman. As far as could be ascertained, all other settlements in that area have changed to the use of watches. Having obtained the agreement of the *falaj* officials in the two areas to study their systems and methods of star use, I drew a halt to the search.

1.5 METHODS OF COLLECTING AND INTERPRETATION OF INFORMATION

The study was carried out over a three year period, formally starting at the beginning of the academic year in October 2005. This was preceded by a preliminary visit, in September 2005, to establish whether such a study would be acceptable to government authorities and academics in Oman, and to establish the extent of support that might be forthcoming. Each year, there were two or three visits of between three and four weeks to Oman, firstly to scope the project and identify the study areas, and later for collecting field data. In the UK and Oman, the available literature and manuscripts were reviewed and experts consulted. Most of the fieldwork and data collection was carried out in the first two years, while the third was used for checking field observations and writing up the findings.

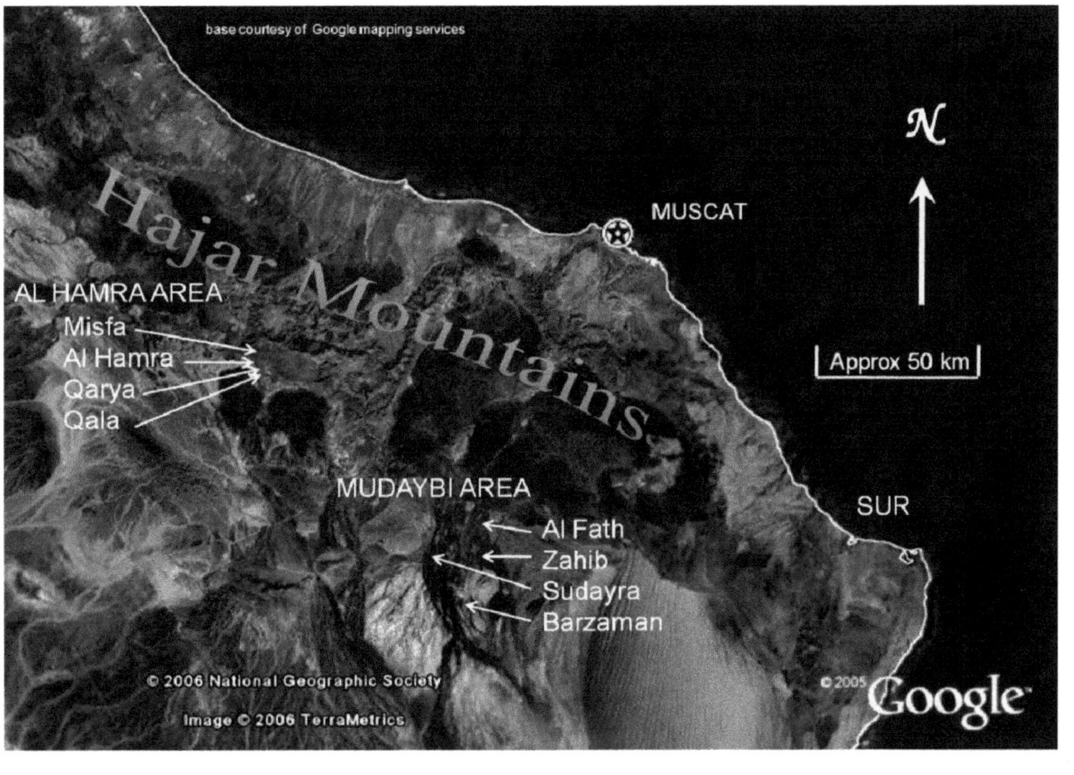

Figure 1.3 Study areas

Primary sources of information used in this study comprise data collected in the field and, rarely, manuscripts written by the *falaj* communities. Secondary sources comprise the available literature on *aflāj*, their construction and organisation, and on astronomy and the use of stars in the region of the Arabian Peninsula. There are few records of telling the time by stars, so the main source of information is data collected in the field, interviewing *falaj* managers and stargazers, and watching the stars with them.

The age of people interviewed was recorded as an indicator of their experience, and because age and failing eyesight have been mentioned in the literature (eg Al-Shaqsi 1996:84) as one reason why stars are no longer used in many places. Although I was not comfortable asking the age of interviewees, Omanis are not shy about their age. However, some people older than c. 50 years were unsure of their age since records of births were not usually made. An extreme example of this is Hamad bin Shinun Al-Hashmi in Sudayra, who, in October 2007, thought he could be in his 50s, 60s or 70s. His wife did not think he was as old as 70, so we agreed on 60s for the purpose of the research.

Times during the study when stars could be watched at night were limited by cloudy skies, the moon when nearly full, and by light and dust pollution. Since it was not possible to predict when the sky would be clear, periods of fieldwork were based on convenience – to avoid the hot summer months and Ramadan as far as possible, when fieldwork would be difficult; and to be present at different times of year to see the different stars.

Introductions to *falaj* officials were usually made by representatives of the MWR, either in person (Figure 1.4) or by telephone. Following the introduction, interviews were usually carried out alone, partly in order not to put too much strain on the resources of the MWR, but also to develop a personal relationship with the villagers and collect and absorb information at my own (slow) pace.

In traditional households, Omani women generally do not meet men outside their immediate family. However, foreign females, unlike males, are free to mix with both men and women in their houses, resulting in a relaxed atmosphere during interviews. The interviews were relatively informal, usually carried out in people's houses or gardens, or at the locations used for stargazing, where the methods of watching the stars were demonstrated to me. I did not use a questionnaire in order to avoid imposing too much control over the flow of discussion and missing critical information. However, after visiting each village two or three times, I made a check list of areas of information needed to obtain a similar scope of information from them. Sound recordings were made mainly indoors when talking about star names and *falaj* organisation. Photographs of people were only taken with the permission of the subject, and all photographs presented in this work are by the author unless otherwise stated.

Many interviews were carried out by day, when it was easier to see any markers used at night and to write down information on star names and timings. Daytime visits were also used to collect information on *falaj* organisation and water ownership. Night visits were most often used to confirm the information given by day, and to identify the stars used. Some were arranged in advance, either to coincide with the informant's time for stargazing or to see a particular star. On other occasions, I visited unannounced to see if anyone was watching the stars and to check on the verbal information given. Several stars were identified using the software program Starry Night™ (Starry Night Pro Plus version 5.8.2 © 2005 Imaginova Canada Ltd.). This shows the night sky

Figure 1.4 Introduction to village elders in Misfa

in real time at any location. The majority of stargazers were able to point out the stars they know on the computer screen, and some looked forward to using it. Any uncertainties were noted and checked again at another time. On my last field visit in March - April 2008, I was loaned a mySky™ GPS star identifier by the Royal Astronomers. This locates the position with GPS and then is used like a gun, pointing at the star and pressing a button; the name of the star is then revealed, provided that there is no interference from metallic objects such as power lines.

The quality and detail of information collected depended to a large extent on the experience, attitude and even health of the people interviewed. I was asked several times whether all of the questions could be given at once, on the assumption that I would then not have to return for further information. After explaining that I needed time to assimilate information, write it up and check for inconsistencies, after which I would probably have more questions, no one complained. However, people's attention often wandered after a while, either because they wanted to be doing something else or because they were bored. For this reason, meetings were generally limited to an hour or less, which meant that it was not always possible to obtain as complete a picture as I would have liked.

Not all meetings were planned. Some chance encounters proved instructive, as when having lunch in the house of a *falaj* official, Saif bin Salim Al-Subhi in Qarya, a farm worker from Al Hamra was also visiting and clarified certain aspects of Al Hamra's *falaj* history and organisation.

The attitude to providing manuscripts varied considerably. Since stargazing is an oral tradition there is little written about it, but I did attempt to see documents of sale and records of *falaj* transactions, such as the auctions of water to raise funds for *falaj* maintenance. Many of the people met were happy for me to see and photograph their documents, but in some places there was a marked reluctance to show personal records relating to property and financial matters. In such cases I did not insist further, in respect of their privacy.

Authors of works on *falaj* organisation and agricultural calendars in Arabia, such as John Wilkinson and Daniel Varisco, were consulted and Paul Kunitzsch provided some valuable information on star names. Use was made of libraries in Oman and the UK to consult published works and theses. Manuscripts related to *falaj* organisation and astronomy held by the Dār al-Makhṭūṭāt (Department of Documents and Manuscripts) in the Ministry of Heritage and Culture in Oman were also viewed.

The main tasks in interpretation of the data collected were firstly to identify the stars, secondly to present and assess the method of stargazing, and thirdly to assess the current status of knowledge and likely continuation of this practice.

Apparently conflicting information was checked by asking the same or similar questions at different times and of different people, and continually checking my own assumptions. This meant that my understanding of the methods and systems used developed during the fieldwork until I was confident that my description is a fair representation of reality. An example of this follows. After gaining some experience of stargazing in Qarya, I was confident in the method used and thought that it would apply in other places. So when two of the village elders in Al Fath said that they watched the top of a tower for the star Zāhara (Capella, α Aurigae), and later my main informant said that different stars were watched in the same location, it became clear that either one of my informants was wrong or that I had not understood. The latter proved to be the case, and both sets of information were correct: Capella is watched as an indicator of the rise above the horizon of other stars.

Identification of the stars used for timing *falaj* water was achieved by looking at the stars, and checking the times of rise and time between them with the information given verbally. The Starry Night™ software was excellent for this purpose. The stars used, the methods of stargazing and the degree of dissemination of knowledge about the system among farmers were compared for the different study villages.

The *falaj* stars and those used in Oman to mark the agricultural seasons as well as for navigation and the stations of the moon of the Islamic calendar were compared to establish the wider context of *falaj* stargazing.

2 OVERVIEW OF THE LITERATURE

This chapter provides a review of the main published and unpublished written sources of particular relevance to the study.

2.1 BOOKS, ARTICLES AND THESES

The main topics addressed, and the sources of information for each, are given in Table 2.1. The term "author" is used loosely, since some of the people named are compilers, editors or translators. However, the exact source of information is given at the relevant point in the text. The two categories of importance are related to the utility for this work only, not to the sources' intrinsic value, which could be much greater if applied to other purposes.

2.1.1 Origin and diffusion of *aflāj*

Wilkinson (eg 1977: 130; 1983) was originally of the opinion that Oman's *aflāj* were introduced from Iran. Writers on the origin and diffusion of these systems around the world are many, and include Goblot (1979), Beaumont et al. (1989) and Lightfoot (2000), generally agreeing that they originated in Iran or central Asia and then spread to other arid and semi-arid areas. Goblot, perhaps, gives the most authoritative account, focusing on Iran where he lived for 20 years. A range of views on the origin and diffusion of these systems is given by Al-Ghafri (2004: 25-26). However, apart from information on when *aflāj* were first used in Oman, this study does not draw on these works.

A key paper by Al-Tikriti (2002) takes account of more recent archaeological excavations than available to the above authors, including his own excavations in the United Arab Emirates. The work of Jocelyn and Jeffery Orchard (2007) is also significant, as they have proposed the earliest date for any *falaj* in the region.

2.1.2 Organisation and management of *aflāj*

There are numerous literature sources describing *falaj* organisation in Oman. Perhaps the best known is Wilkinson (1977), who provides details of the organisation and timing of water distribution based on his fieldwork, probably carried out in the late 1960s. He refers mainly to settlements in the Dakhliyah (Interior) and Sharqiyah (Eastern) Regions, the same districts in which the study areas for this research are located. Wilkinson's descriptions of the responsibilities of *falaj* officials, the distribution of water shares etc. have been much quoted by later authors, and have become almost a standard for comparison. Letts (1978) worked in Oman in the early 1970s and had access to Wilkinson's work during the preparation of her thesis on *aflāj*. Her field area was between Khabura on the Batinah coast and Ibri inland, west of the villages included in this study, and subject to different demographic influences. Letts points out the differences in *falaj* management in her study area compared with that described by Wilkinson. The fieldwork for this research also shows that there are significant differences between villages and that any generalisations must be treated with caution.

Topic	*Aflāj*					Stars				
Author	Origins	Construction and Management	Time keeping	Traditional rulings	Terminology	History of astronomy	Arabic names	Calendars	Navigation	Time keeping
al-ᶜAbrī		B	B				A			
Badr							A			
Evans						A				
Al-Ghafri		B	A		B					B
Goblot	B									
Kunitszch							A			
Lane					B					
Letts		B								
MWR		A								
Selin						A				
Al-Shaqsi		B	B							B
Tibbetts							B		A	
Al-Tikriti	A									
Varisco								A		
Wilkinson	A	A	B	A	B					
Key: A main source; B important source										

Table 2.1 Literature sources

Al-ᶜAbrī (c. 1980) is widely referred to by Omani authors. Although his book was published privately and is out of print, it is available in the library of Sultan Qaboos University. He provides considerably more detail than Wilkinson or Letts on the methods of time keeping, but the most comprehensive treatment of this subject is probably that by Al-Ghafri (2004).

Books and pamphlets published by the Ministry of Water Resources (MWR) are a valuable source of information on water resources (eg MRMEWR 2005) and the distribution and construction of *aflāj* (eg MRMEWR 2002). These are generally based on an inventory carried out by the MWR in the late 1990s. Many of the *aflāj* have been surveyed, and the inventory provides maps showing the access shafts, names of different branches, the point where the village first uses the water for domestic purposes (the *sharīᶜa*, a term explained in more detail in Section 4.3), and the area that is, or could be irrigated. The inventory also provides basic statistics such as the length of tunnel and area of irrigation. This information is stored on the MWR's database, and is referred to as the National Falaj Inventory (NFI).

2.1.3 History of astronomy

The focus of this study is on the current use of stars for a particular purpose, and the history of astronomy is therefore treated only in outline. For this I have used mainly *The History and Practice of Ancient Astronomy* (Evans, 1998) and *Astronomy Across Cultures* (Selin, 2000). Evans provides an accessible overview of the development of astronomy in the Middle East and Mediterranean areas. Although he gives very little information on stars in the Arab world, al-Ṣūfī's work (see Section 2.1.6) is mentioned. The chapters on Egyptian (DeYoung, 2000) and Indian (Kak, 2000) astronomy in "*Astronomy Across Cultures*" have been used for background information.

2.1.4 Time keeping in traditional *falaj* management

Several of the authors listed in Table 2.1 give a description of the use of sundials for time keeping during the day. The most detailed descriptions are given by Al-Shaqsi (1996: 88-91) and Al-Ghafri (2004: 62-66), on the basis that it is necessary to understand the traditional methods of water and land management in order to find ways acceptable to the community to improve the efficiency of water use, which was possibly their main aim.

Information on the use of stars for telling the time in order to distribute *falaj* water is scarce. Wilkinson provides lists of stars used in a few settlements. Al-ᶜAbrī (c. 1980) also gives lists of stars used in several villages and a description from an "old" manuscript of where the stars formerly used in Al Hamra could be seen in the sky and the time that they represented. More recent works include Al-Shaqsi (1996), Al-Siyabi (2000) and Al-Ghafri (2004), all of whom give lists of stars used in one or more villages and, in some cases, the approximate length of time between the stars. Nothing was found in the literature on using stars for timing the allocation of water for irrigation in other countries in the region, and a comparison with other countries was therefore not possible.

2.1.5 Other uses of stars

Several papers have been consulted on the use of stars for purposes other than timing irrigation water. Much has been written about agricultural calendars in the Arabian Peninsula, in which stars are sometimes used to indicate rainy periods and the proper time for activities such as sowing and harvesting. The most prolific author on this topic is probably Daniel Varisco (eg 1994; 1997a; 2000). Star calendars found in mediaeval manuscripts from Yemen are also described by Serjeant (eg 1995c) and others. Possibly more pertinent to Oman is a list of agricultural sayings related to 21 *falaj* stars thought to have been prepared c. 2007 by the Ministry of Agriculture. However the author, date of preparation and villages from which the sayings were collected are not given.

The main source for information on the stars used for navigation in the Gulf and western Indian Ocean is Tibbetts' (1971) translation and comments on a pilot manual written by Ibn Mājid al-Najdī (d. after 906 /1500), a navigator and ship's captain.

King (2000) has written extensively about mathematical astronomy in Islamic civilisation, but this area of knowledge, together with astrolabes, astronomical tables, planetary movements, prayer tables and the determination of the direction of Mecca are peripheral to this study and have not been addressed.

2.1.6 Star names

At the outset of this study, a literature search identified a number of works on Arabic star names. Odeh (1998-2006), for example, has published a list of star names, checked by Paul Kunitszch, on the web. This list, as with many other publications in English, deals mainly with English names derived from Arabic, information which is of limited use for this study.

Much western knowledge about Arabic star names is based on the work of al-Ṣūfī, who used translations (from Greek to Arabic) of, and updated, Ptolemy's Almagest in the 4th/10th C. He also provided Arabic names for stars taken from older texts. A 10th/16th C manuscript copy of this work, *Ṣuwar al-Kawākib al-Thābita* (Illustrated Book of the Fixed Stars) (Al-Sūfī, 995/1586), is held in the Department of Manuscripts and Documents in the Ministry of Heritage and Culture (MHC) in Muscat.

Badr's encyclopaedia (1997 in Arabic) is an invaluable source of information on star names, and has been used extensively to obtain information on star names and their identification. He provides Ptolemaic and modern star and constellation names together with some pre-Islamic names from the Arabian Peninsula, especially those used in navigation on land and at sea. Badr takes information from Ibn Qutayba (d. c. 276/889), al-Ṣūfī (d. 376/986), al-Bīrūnī (d. 440/1048), Kunitzsch and others. Although he comments where his sources differ on the identification of stars or constellations, he does not give citations, which means that the information presented on navigational stars could be from any of his sources.

Paul Kunitszch (eg 1989a; 2004) is renowned for his work on Arabic star names, and provides a more detailed analysis of the origin of certain star names and discussion of their identification than Badr.

2.1.7 Traditional rulings

Rulings on *falaj* building, maintenance and water use in Oman are available from the late 3rd/9th C. Such rulings were written down, not necessarily at the time, and copied by hand or compiled by later authors. Many of these manuscripts have been transcribed and published, mainly by the MHC (eg al-Ḥawārī, 1975; al-Kindī, 1983). Wilkinson (1977: 258-265; 1978) provides the most detailed description of the substance of these rulings. Reviews and translations of similar rulings made in other parts of the Arabian Peninsula have been published in English (eg Ben Shemesh, 1958, 1965, 1969; Maktari, 1971). However, these only apply to the area for which the rulings were made. The laws that pertain in Oman today relate mainly to protection of existing water users, and are summarised by Al-Shaqsi (2004: 155).

None of these sources refers to the timing of water shares, and they are, therefore, not used to any significant extent in this study.

2.2 MANUSCRIPTS, STAR LISTS AND STAR CHARTS

Apart from the work by al-Ṣūfī mentioned above, several manuscripts pertaining to *aflāj* held by the MHC, listed below, were consulted. None of them are from any of the study villages, nor do they mention the use of stars; they were therefore not studied in detail.

- *Kitāb fī taqsīm falaj al-ṣibāra fī wilāyat Adam*: book on the division of water for Falaj al-Sibara in the district of Adam; probably early 1300s/1900s, 33pp. MS 3702;
- *Makhṭūṭa fī falaj al-masmā l-ᶜazīzī bi-balada Sayq*: water rights for Falaj Al-Azizi, Saiq; 1308/1891, 20pp. MS 2316;
- *Falaj Farūḍ*: water rights and transactions for Falaj Farud, Bisya;. 1361/1942 - 1393/1973, c. 19pp. MS 3703;
- *Falaj al-Q.rāḥ*: notebook with water rights for Falaj Al-Q.rah in Bisya; 1353/1934 – 1935, 22 pp. MS 3701;
- *Al-khārija fī masāliḥ al-amwāl*: financial outgoings for Falaj Metha, Bahla; 1387/1967 to 1396/1976.

Few villages in the study area keep written records of *falaj* rights or transactions, but those that were seen are described in the relevant section. Several *falaj* officials provided lists of stars used in their village, and other lists are found in the literature on *aflāj*. A transcription of a manuscript referred to by al-ᶜAbrī (c. 1980: 32-41) provided a list of stars for Al Hamra. A manuscript of water rights for the *falaj* in Mudaybi (1335/1917) was used for *falaj* terms. Two charts of *falaj* stars were located: one in Mintirib and one in Mudaybi, but it appears that they were drawn for the interest of the artists, and were not used by the *falaj* communities.

3 ASTRONOMY AND TIME KEEPING

3.1 INTRODUCTION

This chapter presents a brief introduction to the identification of stars, sufficient it is hoped for the purpose of understanding how stars are recognised. This is followed by an outline of the history of astronomy, particularly in relation to time keeping in the Middle East. To assess the selection of stars used for the *aflāj*, it was considered necessary not only to identify them, but to find out whether they are related to those used for other purposes. Stars used in the region for navigation, agricultural calendars and the stations of the moon of the Islamic calendar are therefore described briefly. The purpose of this section is to provide sufficient background on folk astronomy in the region to be able to assess whether the system of stargazing in Oman for the *aflāj* is related to what is already known.

3.2 IDENTIFICATION OF STARS

Stars may be identified by their:

- position relative to other stars
- location in the sky
- brightness, variability and colour.

The same star may be called different things by different people, so to avoid confusion there is an international classification, usually using a Greek letter followed by the name of the constellation in Latin – in the genitive case – which is unique for each star. Alpha is usually the brightest, beta the second brightest and so on.

3.2.1 Position relative to other stars

The simplest and surest way for an earth-bound observer to identify a star is its position in relation to other known stars. Constellations, describing groupings of stars, are an important tool for star recognition. The constellations are manmade and vary among cultures. The boundaries of the constellations in international use today were set in 1928 by the International Astronomical Union. The sun passes through thirteen of these in the course of the year, and twelve of the thirteen, Ophiuchus being the exception, are the signs of the zodiac used in western astrology:

Capricornus (Capricorn)	the Goat
Aquarius	the Water Carrier
Pisces	the Fish
Aries	the Ram
Taurus	the Bull
Gemini	the Twins
Cancer	the Crab
Leo	the Lion
Virgo	the Virgin
Libra	the Scales
Scorpius (Scorpio)	the Scorpion
Sagittarius	the Archer.

3.2.2 Location

The two main co-ordinate systems used to identify the position of a celestial body outside the solar system are the altitude-azimuth and celestial co-ordinate system. The altitude-azimuth system defines a position by the angle above the horizon (altitude) and direction (azimuth). This is valid only for the time and place of observation, but is a useful tool for knowing where to look for a particular object and for recording what is seen in the field.

The celestial co-ordinate system is the parallel of the Earth's co-ordinate system but for the celestial "sphere" surrounding the Earth. The celestial equator, north pole and south pole are the projection of the Earth's equator and poles onto the celestial sphere. The celestial meridian (i.e. 0° longitude) runs between the poles through the vernal equinox, which is the point where the celestial equator meets the ecliptic. The ecliptic is the Earth's path around the sun, and the sun's apparent path around the Earth during the course of a year, hence the vernal equinox is the time when the sun crosses the celestial equator in March. In this co-ordinate system, the location of an object is defined by its declination, equivalent to latitude on Earth, and right ascension, which is measured in hours from the meridian, equivalent to longitude. The system is illustrated in Figure 3.1.

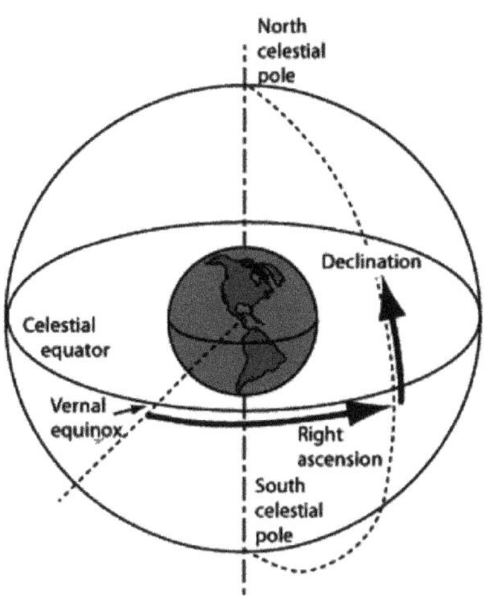

Figure 3.1 The celestial sphere
from Nave (2007), with permission

3.2.3 Brightness, variability and colour

The brightness of a star is described by its magnitude according to a system introduced by Hipparchus c. 150 BC (Mosley, 2004: 19). The brighter the star, the lower the magnitude: a star of magnitude 1 is 2.512 times brighter than magnitude 2, which is 2.512 times brighter than magnitude 3, and so on. Sirius (α Canis Majoris), the brightest star, has a magnitude of c. -1.4. The human eye can detect objects down to c. magnitude 6. The brightness of some stars varies perceptibly: the magnitude of Algol (β Persii) and Betelgeuse (α Orionis), for example, varies from 2.2 to 3.47 and 0.4 to 1.3 respectively.

The colour of a star is dependent on its surface temperature, as shown in Table 3.1 (Menzel, 1975: 355). While it is difficult for the untrained eye to distinguish each of these classes, it is relatively easy to see the extremes. For example, the very blue to blue stars cannot easily be confused with red stars unless you are colour blind.

Colour	Surface temperature, K	Example
very blue	50,000	ζ Puppis
blue	25,000	α Leonis (Regulus)
green	11,000	α Canis Majoris (Sirius)
white	7600	α Canis Minoris (Procyon)
yellow	6000	α Aurigae (Capella)
orange	5100	α Tauri (Aldebaran)
red	3000 – 3600	α Orionis (Betelgeuse)

Table 3.1 Relationship of star colour to surface temperature

3.3 STARS AND TIME

The distance between different stars can be used to estimate spaces of time: the stars rise approximately four minutes earlier each night so different stars are visible at different times of year, but the time between them remains the same. According to the Holy Qur'ān (Sūrat al-Anᶜām, VI: 96-97), the sun and moon were given by God for the reckoning of time, and the stars to guide man on land and at sea (Ali, 1968: 317). There is no mention in the Qur'ān of the use of stars for telling the time, but it appears that celestial bodies have been seen as constellations and used by man for some form of time reckoning since the earliest times.

Kak (2000: 312) names some of the constellations recognised in India and describes a centennial calendar with a cycle of 2700 years based on the relative position of Ursa Major and the Pleiades, possibly as early as the 7th millennium BC. Similar constellations were known in ancient Sumer, Egypt and Babylon, and later by the Greeks. It appears that all of these ancient civilisations used a lunar-solar calendar, and a lunar calendar is described in the texts of the Rig Veda, in which the path of the moon was marked by stars, a precursor to the modern Islamic stations of the moon. Kak (2000: 307-308) argues that the period that the calendar refers to is before 1900 BC and possibly before 2600 BC.

In ancient Egypt, star markers were used to divide time into 10-day periods, and the heliacal (dawn) rising of Sirius (α Canis Majoris) was associated with the onset of the Nile flood from the First Dynasty (3000-2840 BC). Star calendars or clocks for telling the time at night, with approximately one hour between the stars, have been found in coffins dating from approximately 2106-1068 BC. They were designed to identify the last hour before dawn for the purpose of prayer or worship, and telling the time was the responsibility of hour-watching priests (DeYoung, 2000: 481-486). The stars used for this purpose have not been identified.

The Babylonian creation theory, the *Enuma Elish*, describes a star calendar that may have existed before 1600 BC, with three stars to each of the twelve months of the year (Evans, 1998: 10). The astronomical text, the MUL.APIN and clay tablets with diagrams of star lists from the period c. 650 BC are described in Evans (1998: 5-11). The MUL.APIN provides a star calendar, with lists of stars that set as others are rising and sayings relating the behaviour of celestial bodies to floods and harvests. As in the *Enuma Elish*, three stars marked each month, apparently one from near the celestial equator, one north and one south of the equator. This is illustrated in Evans (1998: 10 Figure 1.3), together with the identification of many of the stars and constellations. Several planets were also used. However, the identifications given do not correspond to three sets of star markers at a similar latitude and, as Evans points out, the use of planets makes the calendar of questionable use for time reckoning. Day and night were each divided into three watches using water clocks by day, but it is not stated whether water clocks or stars were used at night.

Also written in about 650 BC, Hesiod's *Works and Days* provided an agricultural calendar in which stars play a small part. This was the first of many Greek texts relating stars to agriculture and weather, of which perhaps the best known is the *Phaenomena* of Aratus, written in c. 250 BC. However, there is no evidence that stars were used for telling the time. In c. 150 BC Ptolemy described 48 constellations, which are much the same as today's, and the stars forming them. Mosley (2004: 138-140) considers that 30 of those constellations derive from Babylonia, and that many have earlier roots in Sumeria.

Muslim astronomers drew on Greek knowledge and several translations to Arabic of Ptolemy's work, known as the *Almagest*, were made in the 3rd/9th C. Al-Ṣūfī (d.376/986) used these translations and updated the *Almagest* with his own observations in his *Ṣuwar al-Kawākib al-Thābita* (Illustrated Book of the Fixed Stars), and later Muslim astronomers such as al-Bīrūnī (d. 440/1048) drew on his work (Kunitzsch, 1989b). Al-Ṣūfī also provided older Arabic names for stars, from earlier Arabic texts and refers to the work of Abū Ḥanīfa l-Dīnawarī (d. c. 282/895) (Schjellerup, 1874: 42) as the most reliable of them. In about 1000 AD, the *Almagest* was translated from Arabic into Latin and reintroduced to Europe via Spain, forming the basis of modern western astronomy and bringing many Arabic star names to the west.

3.4 STARS FOR NAVIGATION

Stars would have been used for navigation on land and sea in the region from the earliest times. Agius (2008: 277-278) argues that pre-Islamic desert poets were familiar with ships and the sea, and it is therefore possible that the early settlers in northern Oman were familiar with some of the stars used at sea, as well as those that they used on land. There is therefore a possibility that some of the stars used for *falaj* timing were selected from navigational stars. The use for navigation is related to time keeping as well as to direction, in that all except circumpolar stars indicate a different direction at different times between rising and setting.

3.4.1 Navigation on land

In relation to the Arabian Peninsula, Badr (1988: 3) comments on the predilection of the Arabs to see stars in pairs, using them to know their location in the wide desert, without which knowledge they would be led to death and destruction. Indeed, the common use of pairs of stars for *falaj* purposes was noted in the early stages of my fieldwork, and the clear application to navigation rather than to time keeping, before reading Badr's work. Unless a star is very far south or north, such as Canopus (α Carinae) or Polaris (α Ursae Minoris) respectively, its direction relative to an observer changes during the night. Orion's sword, for example, points south; but if only one star could be seen, the observer would not know the direction. Thus it is usually necessary to use at least two stars to know the direction.

Nowadays there is little need for stars for desert navigation: most people travel by day and have vehicular transport, using tracks and roads with known destinations. A Bedu family in Sudayra, when asked about their former use of stars during the fieldwork for this study (November 2006), mentioned the Pole Star, Yēh (CA Jāh), both for navigation and for its use in pre-Islamic times as the direction in which the head of a corpse was placed for burial. However, as a circumpolar star that does not rise or set, it is of little use in time keeping. Shābik and Thurayyā were also mentioned as stars that move from place to place in the sky, in contrast to the Pole Star, and Kuwī as a star that brings very cold weather.

The importance of the Pole Star for navigation on land is illustrated in the following passage from the story of the journey of two Jesuit captives from Dhofar to Sanᶜā via Ma'rib in 1590 (Beckingham and Serjeant, 1950: 201), referring to their Muslim captors:

> "... they lingered very little; there is not a drop of water in all that desert and so they were very afraid that the weather might become cloudy and prevent them from seeing the Pole Star, by which they were guided at night. There was no path or sign of one, and if they had not travelled at night the water they carried would not have sufficed."

3.4.2 Navigation at sea

A major source of information on the use of stars for navigation at sea is Tibbetts' (1971) translation and commentary on a pilot manual for the Indian Ocean compiled by Aḥmad b. Mājid (Ibn Mājid, d. after 906/1500). The star compass given in this work (Tibbetts, 1971: 297) has been redrawn to show star identifications in addition to the Arabic name (Figure 3.2). There are 16 star markers in all, one of which is the Pole Star (α Ursae Minoris). The others are shown twice: rising to the east and setting to the west as a mirror image, with the Arabic and transliteration given on the east and the formal identification on the west. This is illustrated by the blue line: ᶜAqrab is α Scorpii, rising in the southeast and setting in the southwest.

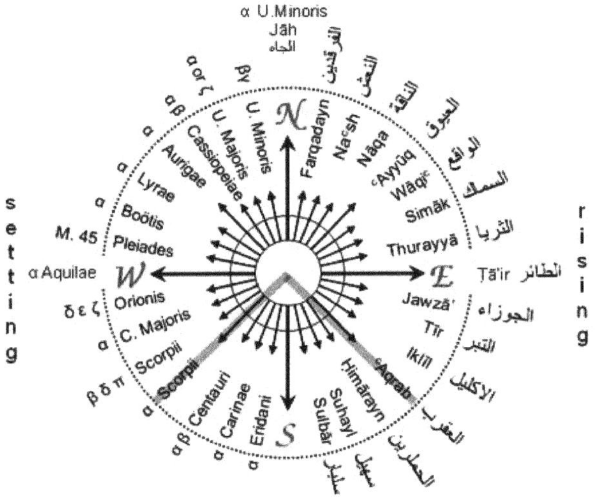

Figure 3.2 Navigational stars in the Indian Ocean
after Tibbetts (1971: 297)
C. = Canis, M. = Messier object, U. = Ursae

These stars are only meant to indicate the direction, not to be exact: indeed, α and β Centauri (al-Ḥimārayn) rise and set south of α Eridani (Sulbār). However, these cannot all be seen together in the sky, at least from the latitude of northern Oman, and any of them can be taken as a rough indication of S or SSE/SSW.

Additional stars would have been, and probably still are, used to cover all points of the compass and all nights during the sailing season. Serjeant (1995a) and Agius (2005: 155-173; 2008: 196-199) mention the stars used by Arabs and Iranians for navigation at sea. Serjeant focuses on the translation of texts, and Agius on the tools and methods used, for example for measuring latitude. Several Arab and Indian star compasses used in the Indian Ocean are described by Arunachalam (1996: 270-273), but identifications are only given for a few of the stars; one such compass is given for the Batinah coast of Oman by Donaldson (1978: 158), with some local variations on the names in Figure 3.2, including (al-) Layā' for Polaris. However the literature on this topic is sparse and it appears that many stars used for navigation in the region have not been identified or recorded.

3.5 STARS IN AGRICULTURAL ACTIVITIES

Certain stars have been associated with agriculture for millennia. Hesiod's *Works and Days*, written in c. 650 BC is often quoted as an early example of the use of stars in agriculture in Greece, although it by no means represents a calendar for a full year and many other indicators are used. Table 3.2 gives references to stars in the translation by Dorothea Wender (Hesiod, 1973).

Varisco (eg 1997a) has described traditional star calendars for Arabia using 28 star markers for rain, agricultural activities and other events. These are recorded in the works of several 4th/10th to 5th/11th C writers, in which the star markers, especially for rain, are known as *anwā'* (s. *naw'*). He (eg 2000: 622, 625-626) argues that the original pre-Islamic *anwā'* were fewer than 28 but were later modified in Islamic times to incorporate the stations of the moon, introduced from India. Elsewhere, he gives a list of the earlier *anwā'* stars according to Abū Zayd al-Anṣārī (d.206/821) and Quṭrub (d. 215/830), with his own tentative identifications shown in Table 3.3.

Varisco has also published numerous articles and a book on almanacs from medieval Yemen (eg Varisco, 1994, 1997a). Further examples come from Dhofar (Alian, 2006: 15-16), Socotra (Serjeant, 1995b) and Qatar (Varisco, 1990). It appears that the compilation of such almanacs was mainly a pursuit of the learned, and that they were by no means a practical guide to the seasons.

Star	Action/effect of star	Time of year	Activity
Arcturus	evening rise – sixty days following the solstice	February - March, preceding the swallow	harbinger of spring – prune vines before the swallow comes
Pleiades	reappear after forty days		sharpen sickles
Pleiades	rise (morning)	early May	harvest (winter wheat)
Pleiades	snails climb up plants to avoid them	mid-May	stop digging vineyards - sharpen sickles and start the harvest.
Orion	rises (morning)	July	winnow the grain
Orion and Sirius	move into the mid-sky	September	pick grapes
Arcturus	morning rise		
Sirius	passes overhead less in day, more at night	with other things – autumn + rain (October)	good time for cutting timber (+ making tools, eg ploughs)
Pleiades, Hyades + Orion	sink	the old year dies. October	plough
Pleiades pursued by Orion	sink (plunge into the clouded sea)		gales – do not sail
Pleiades	set (morning)	November	plough
Pleiades, Hyades + Orion	set (morning)		sow

Table 3.2 Stars in Hesiod's *Works and Days*

***Anwā'* star name**	**Identity**	**Transliteration to MSA**	**English name**
from Varisco (1997b: 255)			
al-ᶜArquwatān al-Mu'akhkharatān	α + β Pegasi	العرقوتان المؤخرتان	Markab + Scheat
Sharaṭ/Sharaṭān	β + γ Arietis	شرط/شرطان	Sheratan + Mesarthim
Thurayyā	Pleiades	ثريا	Pleiades
Jawzā'	parts of Gemini and Orion	جوزاء	
Dhirāᶜ/Dhirāᶜān	α + β Geminorum	ذراع/ذراعان	Castor + Pollux
Nathra	ε, γ + δ Cancri	نثرة	Asellus Australis (δ)
Jabha	ζ, γ, η + α Leonis	جبهة	Regulus (α)
ᶜAwwā'	β, η, γ, δ + ε Virginis	عواء	Vindemiatrix (ε)
Ṣarfa	β Leonis	صرفة	Denebola
Simākān	α Virginis + α Boötis	سماكان	Spica + Arcturus
Nasrān	β, γ + α Aquilae α Lyrae	نسران	Altair (α Aquilae) Vega (α Lyrae)
Akhḍar		اخضر	
al-ᶜArquwatān al-Ūlayān	δ + γ Pegasi	العرقوتان الوليان	δ Pegasi = α Andromedae = Alpheratz; γ Pegasi = Algenib

Table 3.3 *Anwā'* star markers

Other star calendars from the Arabian Peninsula referred to in the literature include a Canopus (α Carinae) calendar of the Gulf region, described by Varisco (1990: 11-14; 2000: 629) and mentioned by Wilkinson (1977: 11), and a Pleiades - lunar calendar from Yemen (Varisco, 2000: 630-633). The Canopus calendar comprises 10-day periods following the dawn rising of Canopus in late summer and covering the seasons of autumn and winter. The Pleiades calendar is, or was, based on its conjunction with the moon, starting at the end of September, 19 days after the new moon. Each period in the calendar is two days shorter than the previous, ending with the new moon in April, when the Pleiades disappear from view. These winter calendars are clearly of more practical use for agriculture in the Middle East than the somewhat artificial full-year almanacs.

A list of Omani agricultural sayings related to stars was provided by the Astronomers of the Royal Court, Mohammed Salim Al-Busaidi and Suleiman Hallal Al-Busaidi in March 2008. The author and sources of information are not given, although the document is modern, having been prepared on a word processor, and is thought by several people to have been compiled by the Ministry of Agriculture. It lists 21 stars by the same names, and with similar information on timings as those in the Qarya area (see Chapter 6). It gives the number of days and dates for each star, a list of agricultural activities for each period and examples of star lore, often rhyming, for the majority of stars, two of which are given in the discussion of the selection of stars in Section 11.5.

3.6 STATIONS OF THE MOON

The stations of the moon comprise 28 stars (Table 3.4), probably introduced to Arabia from India in pre-Islamic times and later incorporated into the Islamic calendar (Varisco, 2000: 622). It seems that they were, at least in theory, selected as stars close to where the moon is seen on consecutive nights of the month, and they are therefore at roughly equal time intervals and aligned on the ecliptic. However, there is no evidence that they were ever used for time keeping in the Islamic world.

Arabic (MSA) star name	Identity	Transliteration of MSA	English names of the brighter stars
from Badr (1988: 200-205)			
الشرطان	α, β + γ Arietis	al-Sharaṭān	Hamal (α), Sheratan (β)
البطين	ε, δ + ρ Arietis	al-Buṭayn	Botein (δ)
الثريا	Messier 45	al-Thurayyā	Pleiades
الدبران	α Tauri	al-Dabarān	Aldebaran
الهقعة	λ, $\varphi^1 + \varphi^2$ Orionis	al-Haqʿa	Meissa (λ)
الهنعة	γ + ξ Geminorum	al-Hanʿa	Alhena (γ)
الذراع	α + β Geminorum	al-Dhirāʿ	Castor (α), Pollux (β)
النثرة	ε, γ + δ Cancri	al-Nathra	Asellus Australis (δ)
الطرف	λ Leonis + κ Cancri	al-Ṭaraf	Altarf (λ Leonis)
الجبهة	ζ, γ, η + α Leonis	al-Jabha	Regulus (α)
الزبرة	δ + θ Leonis	al-Zubra	Zosma (δ)
الصرفة	β Leonis	al-Ṣarfa	Denebola
العواء	β, η, γ, δ + ε Virginis	al-ʿAwwāʾ	Vindemiatrix (ε)
السماك	α Virginis	al-Simāk	Spica
الغفر	τ, κ + λ Virginis	al-Ghafr	
الزباني	α + β Librae	al-Zubānī	Zubenelgenubi (α) + Zubeneschimali (β)
الاكليل	β, δ + π Scorpii	al-Iklīl	Graffias (β), Dschubba (δ)
القلب	α Scorpii	al-Qalb	Antares
الشولة	λ + ε Scorpii	al-Shawla	Shaula (λ)
النعائم	γ, δ, ε, η , σ, φ, ς + ζ Sagittarii	al-Naʿā'im	Kaus Australis (ε)
البلدة	no star	al-Balda	
سعد الذابح	α + β Capricorni	Saʿd al-Dhābiḥ	Algedi (α), Dabih (β)
سعد بلع	ε, μ + ν Aquarii	Saʿd Bulaʿ	Al Bali (ε)
سعد السعود	c^1 Capricorni, β + ξ Aquarii	Saʿd al-Suʿūd	Sadal Suud (β Aquarii)
سعد الاخبية	γ, π + η Aquarii	Saʿd al-Akhbiyya	Sadachbia (γ)
الفرغ الاول	α + β Pegasi	al-Fargh al-Awwal	Markab (α), Scheat (β)
الفرغ الثاني	α Andromedae + γ Pegasi	al-Fargh al-Thānī	Alpheratz (α And.) + Algenib (γ Peg.)
بطن الحوت	β Cassiopeiae	Baṭn al-Ḥūt	Caph

Table 3.4 Stations of the moon

4 THE *FALAJ* SYSTEM IN OMAN

This chapter provides a brief description of the different types of *falaj*, followed by an outline of the history of their construction in Oman, the use of *falaj* water for different purposes, particularly for agriculture, and the traditional methods of water management, within which the use of stars plays a small but possibly significant role. It is not intended to provide a comprehensive picture of *falaj* systems, only to set the picture for the information presented for the study villages in subsequent chapters. For further detail, the reader is referred to the authors listed in Table 2.1.

4.1 *AFLĀJ* AND THEIR DIFFERENT TYPES

In Classical Arabic, the term *falaj*, pl. *aflāj*, has many meanings, including to divide into shares and running water (Lane, 1968 vi: 2426), which relate to its present day use in Oman where it is applied to the whole system of tapping groundwater or wadi flow, transporting it and distributing it to fields by gravity flow. This term has been in use for a long time: it appears in al-Ḥawārī (fl. late 3rd/9th to early 4th/10th C) (1975 ii: 212-228).

In the *Encyclopaedia of Islam* New Edition (Index Volume Fascicule 2 Glossary and Index of Terms, 2006: 235), *falaj* is described as:

> the term used in Oman, Trucial Oman and Bahrain to designate an underground aqueduct with surface apertures to facilitate cleaning. This type of aqueduct, which may be of Persian origin, is now called SĀḲĪ (pronounced *sāḏjī*, pl. *sawāḏjī*) in al-Aflāḏj, the district in Naḏjd which takes its name from ~.

This definition is only partially correct for Oman, referring to only one type of *falaj* (the *dā'ūdī* type, see below) and only one part of the *falaj* system, that for conducting water from its source to the fields. However, the information given helps to define the area in which this term was and is used.

An inventory carried out by the Ministry of Water Resources (MWR) in the late 1990s found c. 4000 *falaj* systems, of which c. 3000 were still in use. Their approximate locations are shown in Figure 4.1: they are most commonly found in the north of Oman, in the foothills and low lands bordering the northern mountains. The MWR (eg MRMEWR, 2002: 7) recognises three types of *falaj*: *ghaylī*, *ᶜaynī* and *dā'ūdī*. In all three, the hydraulic principle is similar, in that water is conveyed at a shallower gradient than it would flow naturally, and thus the land can be cultivated by gravity flow without having to lift the water either from the wadi or from the water table.

The ultimate source of water for the *falaj* systems is rainfall within and near the mountains. Rainfall can directly recharge the aquifers within the mountains, mainly limestones and ophiolites, and it also runs off forming wadi flow, often only lasting for a few hours. The wadi gravels, the main aquifer tapped by *aflāj* away from the mountains, are recharged from wadi flow and from seepage from the limestones and ophiolites. Water in the limestones and ophiolites is largely stored in and transmitted by fractures. The limestones are karstified, with numerous caves and give rise to some large springs. The ophiolites feed smaller, flashier springs: although they do not form what can be termed a major aquifer, they are present over a wide area and provide important local resources. The wadi gravels store water in and transmit it through pore spaces. In places they are cemented, and the water follows preferential flowpaths in the less cemented zones.

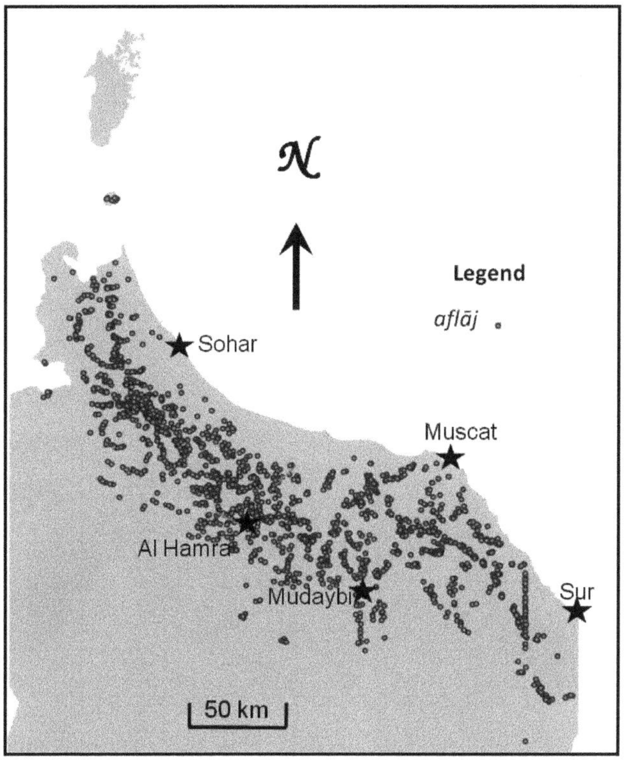

Figure 4.1 Locations of *aflāj* in Oman

A *ghaylī falaj* (Figure 4.2 a) draws water from wadi flow, either at or just below the surface of the wadi bed. Sluices are built to divert the flow, which may be channelled several kilometres to the area to be irrigated. The flow is highly variable, depending on recent rainfall and runoff, and sometimes stops altogether. This type of *falaj* is therefore used mainly for seasonal crops, such as wheat and cotton. In the villages considered in this study, there are no *ghaylī falaj* systems.

An *ᶜaynī falaj* (Figure 4.2 b) draws water from one or more springs, which occur where groundwater meets the surface. There are many springs from the limestones and ophiolites in the Hajar Mountains which feed *ᶜaynī aflāj*.

As for *ghaylī aflāj*, the water may be transported some distance before use and the base and sides of the channel are often sealed with cement to limit losses from seepage. A *dā'ūdī falaj* (Figure 4.2 c) like an *ᶜaynī falaj* taps groundwater. The main components are:

- shafts connected beneath the water table in the source area. The source is usually explored first to ensure that the aquifer can supply the system;
- surface channels: usually a single channel through the main settlement after which it is divided to reach the cultivated areas;
- one or more tunnels dug back into the watertable, with access shafts for removal of spoil and for ventilation; usually there is one main tunnel, but tributaries with subsidiary sources may also be built to provide sufficient flow.

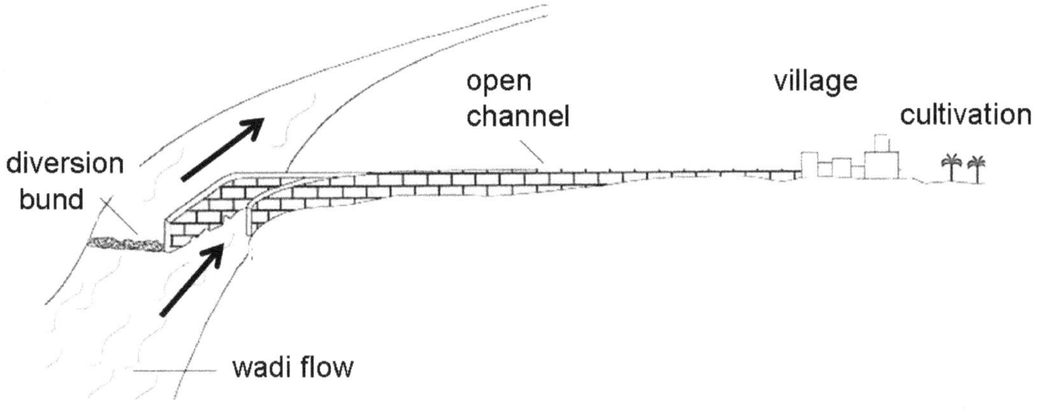

a) Ghaylī

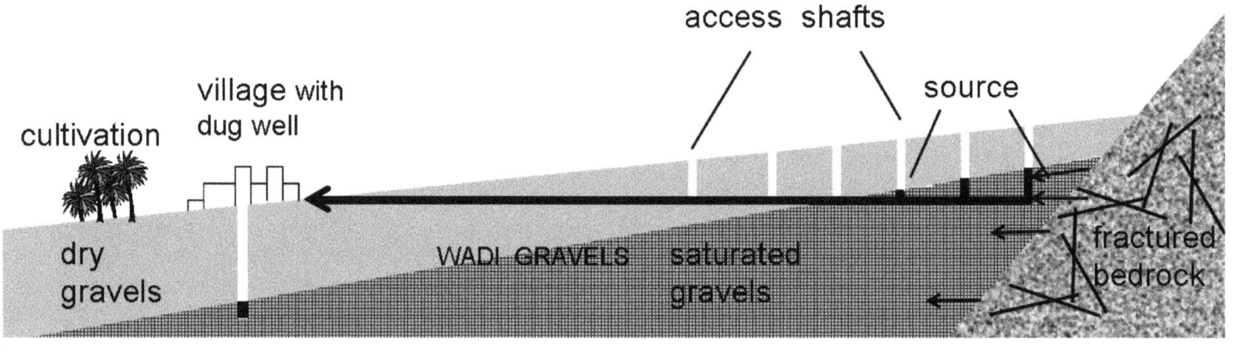

b) ᶜAynī

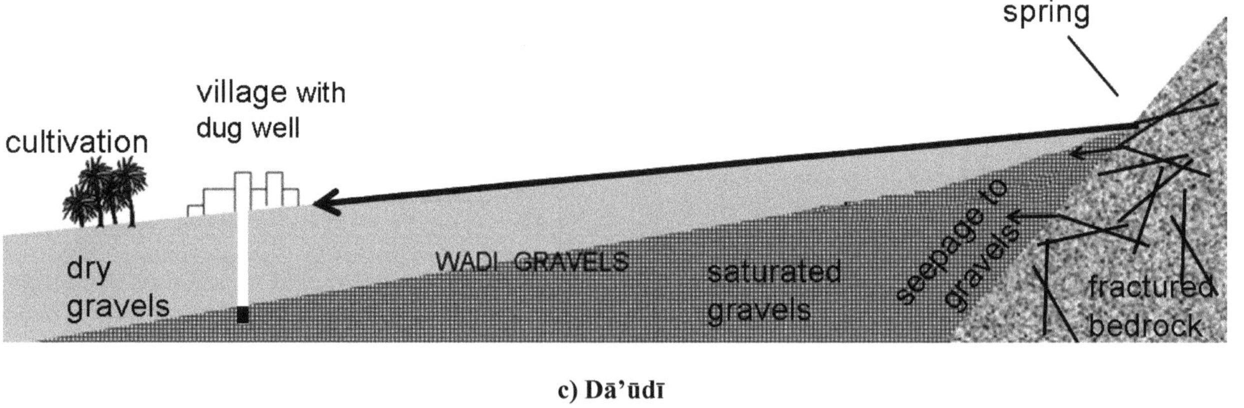

c) Dā'ūdī

Figure 4.2 ***Falaj*** **types**
redrawn from MRMEWR (2005:40 Figure 3)

The source area acts like a collector well, providing a large surface area for groundwater to flow through into the tunnel, and also providing more storage than that present in the wadi gravels prior to excavation.

The *dā'ūdī falaj* is of the same design as the qanat (MSA *qanāt*, pl. *qanawāt*) systems of Iran, which are so numerous (Iran has over 30,000 such systems still running) and have been so widely written about, eg Goblot (1979), Beaumont et al. (1989), that the term *qanāt* has been applied to these systems in Oman, possibly first by Wilkinson (1974; 1977). However, in Oman, *qanāt* means a channel or canal as in its usual usage in MSA, and is sometimes used for the tunnel transporting water in the *dā'ūdī falaj*, but is not applied to the *falaj* system as a whole. This type of irrigation system is widespread in arid and semi-arid areas of the world with wetter hinterlands that provide groundwater recharge (see Goblot, cited above, and Lightfoot (2000) for example).

The *ᶜaynī* and *dā'ūdī* types of *aflāj*, which tap groundwater, have a more reliable yield than the *ghaylī aflāj*, which depend on surface or near surface flow. All three types require regular maintenance of the channels for the removal of silt and plant roots. Occasional floods can cause considerable damage, and repairs are most difficult and costly on the tunnel sections of *dā'ūdī aflāj*. However, modern repairs often use concrete to line the tunnels and shafts, at least outside the source area, potentially reducing any losses from seepage from the tunnel to the unsaturated gravels. This, and the protection of the tops of shafts with concrete lids, also reduces the likelihood of significant damage occurring in the future.

4.2 HISTORY OF *FALAJ* CONSTRUCTION

It is reasonable to assume that springs, i.e. *ᶜaynī aflāj*, have been used for irrigation as long as there have been agricultural settlements in the region, since at least the 3rd millennium BC (eg Cleuziou and Tosi, 2007: 139-151), and the use of water from the wadis, i.e. *ghaylī aflāj*, requires little additional engineering expertise. The following discussion of the origins of *aflāj* therefore applies mainly to the *dā'ūdī* type, which require considerable expertise both for their construction and their maintenance.

It appears that such *aflāj* were first constructed in Oman (and adjacent countries in the Arabian Peninsula) when drier conditions, and possibly an increasing population, made irrigation necessary on a larger scale than that feasible with dug wells. In legend, the *dā'ūdī aflāj* were created by the Prophet Sulaymān b. Dā'ūd (Solomon, son of David), who is thought to have lived in the 10th century BC. He is said to have stayed in Oman for ten days and, because the land was so dry, ordered the *jinn* (genies) to dig 1000 such *aflāj* every day. This explains why this type of *falaj* is called "*dā'ūdī*" (Wilkinson, 1977: 130).

Several authors, especially the earlier writers such as Wilkinson and Goblot, have concluded that the technique of *falaj* construction was introduced from Persia by the Achaemenids (c. 550 – 330 BC). However, Lightfoot (2000: 224) refers to the possibility that a few *aflāj* were built in Oman before Achaemenid times, and at least six *falaj* systems dated to c. 1000 BC (Iron Age) have been identified in the United Arab Emirates and Oman (Al-Tikriti, 2002: 137).

Al-Tikriti proposes that Iron Age *aflāj* in Oman and the United Arab Emirates are of a pre-Achaemenid local origin. Regardless of the place and time that these systems were developed, his argument that they were built in Oman and adjacent countries in the Arabian Peninsula before Achaemenid times is convincing. It is also consistent with the legend of King Solomon. It therefore appears that irrigation was generally necessary throughout this region in the Iron Age and that the larger agricultural settlements of this period would have depended on *falaj* systems.

Jocelyn and Jeffery Orchard have published works on their excavations of Bronze Age (3rd millennium BC) settlements in northern Oman (eg Orchard, 1995), indicating an even earlier origin of *aflāj*, and one *falaj* system excavated by them near Bahla is attributed to c. 3000 BC (Orchard and Orchard, 2007: 150 - 151), during the Bronze Age. Since numerous Bronze Age settlements were dependent on agriculture, *aflāj* would have been needed if the climate were too dry for rain-fed agriculture and the population too large for wells to supply sufficient irrigation water. Jorgensen and Al-Tikriti (2002, 2nd author incorrectly given as al-Tikiriti) present evidence for increasing aridity and a decline of groundwater levels from c. 2500 BC in the area of Al Ayn, UAE, but although several Bronze Age wells have been found, so far there is no evidence of any Bronze Age *aflāj* in that area[2].

According to Wilkinson (1977: 130), there were major developments of the *falaj* network in Oman during Achaemenid times and in the Sāsānian period (mid-3rd to mid-7th centuries AD). His timing of the first period may be at least partly due to the assumption that the *aflāj* were introduced by the Achaemenids, but Goblot (1979: 56) mentions, without giving his source, that Cyrus the Great (d. 530 BC) and Cambyses II (d. 522 BC) ordered *aflāj* to be built in Oman. Wilkinson (1977: 50) estimated that the area of land irrigated by *aflāj* was at its maximum in the late Sāsānian period. There was another major period of construction and repair following widespread destruction of *aflāj* during a period of civil war, from the mid-11th/mid-17th to mid-12th/mid-18th C during the rule of the Yaᶜāriba Imams, when Oman was politically stable (Wilkinson 1977: 50).

Although the origin of *aflāj* in Oman is uncertain, both in time and place, it appears that these systems have

---

2 Walid Al-Tikriti, personal communication, 21 November 2006.

provided communities in this area with water for irrigation and domestic purposes for at least 3000 years. As archaeological work in the area slowly progresses, earlier and earlier origins are being established. It remains for the archaeologists, and those researching climate change and the development of agriculture in the area to establish the origins of these systems.

Although it is now too expensive to construct new *aflāj*, several are of relatively recent construction, such as those in three satellite villages to Mudaybi (Zahib, Ridda and Al Shariq) which were built in the early 20th C.

4.3 *FALAJ* WATER USE

4.3.1 The principle of sharing *falaj* water

There are many references to water in the Holy Qur'ān, especially to rivers flowing beneath gardens for the righteous in paradise (eg Sūrat al-Baqara, II: 23; Sūrat Āl ᶜImrān III: 198); and rain (eg Sūrat al-Baqara, II: 164) and springs (eg Sūrat Yā Sīn: XXXVI: 34) given by God for watering animals and growing plants. Only a few verses refer to water rights, but the principle of sharing water is clear: the following are taken from a translation into English by Ali (1968):

Sūrat al-Shuᶜarā'
XXVI: 155
He said: Here is a she-camel: she has
A right of watering.
And ye have a right
Of watering, (severally)
On a day appointed.

Sūrat al-Qamar
LIV: 27-28
For We will send
The she-camel
By way of trial for them ...
And tell them that
The water is to be
Divided between them:
Each one's right to drink
Being brought forward
(By suitable turns).

The Ḥadīths are sayings and deeds attributed to the Prophet that bear considerable weight in Islamic law. They give more specific guidance on how water should be used, in particular on the amount and sequence of irrigation. One Ḥadīth quoted by many authors explaining water rights (eg Wilkinson, 1974: 19) is that land should be watered to a certain depth in sequence, from upstream to downstream. But whatever the Qur'ān and Ḥadīths say, it appears that any arrangement for sharing water is acceptable provided that it is agreed by all parties (Wilkinson, 1977: 102).

Fiqh is defined in the *Encyclopaedia of Islam* New Edition (Index Volume Fascicule 2 Glossary and Index of Terms: 242) as "the technical term for jurisprudence, the science of religious law in Islam". In Oman, decisions made by the *ᶜUlamā'* (Muslim scholars of religion) have been written down from early times in collections of decisions or advice on various aspects of interpretation of Islamic law. These works generally comprise the posing of a problem or question, followed by the authority's opinion on it. One of the earliest works referring specifically to *aflāj* is the collection of al-Ḥawārī (d. early 4th/10th C), which has been transcribed and published by the MHC (al-Ḥawārī, 1975). Wilkinson (1977: 258-265; 1978) summarises the substance of these rulings, which are mainly related to problems such as water rights and *falaj* maintenance: they do not address the practicalities of timing the water shares.

4.3.2 Order of water use

Traditionally, the *aflāj* provided all of the water needs of the agricultural settlements that they served, and in many places there was no other source of water. An outline of the order of *falaj* water use is given in Figure 4.3. For *dā'ūdī aflāj*, the first possible use of the water is from the access shafts: if these are left open they are free for anyone, not only the *falaj* community, to use for drinking or watering animals. However, the access shafts are rarely used these days, as most travellers buy water from pumped wells if needed.[3]

Water from a *dā'ūdī falaj* usually reaches the surface just above the main settlement. Drinking water was taken here from the main channel, the point known as the *sharīᶜa*, first for humans, then animals, followed by ablutions at the mosques, baths, and a place for washing corpses, not always in that order except perhaps for the *sharīᶜa*, which was usually located above other points of use to provide clean water. For *ᶜaynī aflāj*, the *sharīᶜa* is located at a convenient point upstream of the settlement, and the type and order of water use is the same as for *dā'ūdī aflāj*. If the settlement was large and the *sharīᶜa* therefore too far (more than 1 km, say) to provide ready access, wells would have been used for drinking water, as in Bahla.[4]

The *sharīᶜa* is described by Wilkinson as the point: where drinking water may be drawn: of first permitted use of water; of free access to all (1974: 13); and later as the point where pure water is drawn (1977: 267). It is defined in Wehr (1966: 466) as "a source, water hole, drinking place", and by Badger as "From which to draw water, of a river" (1881: 1205). Its other meaning is Islamic Law, and the two uses are clearly linked by the concept of what is lawful, and also perhaps by the concept of purity.

Apart from drinking water for humans, *falaj* water is still widely used for all of the domestic purposes mentioned above, several of which are illustrated in the sections on the study villages. There are often additional places for bathing and for washing dishes and clothes within the irrigated area.

Downstream of the first set of baths and mosques, often one of each for men and women separately, the channel is divided and subdivided so that water can reach all of the cultivated land. Movable dams, of stone, metal, cloth, wood and, more recently, concrete, are used to block some channels while others are running, so that the land

[3] Ali Al-Yahyai, personal communication, 8 March 2005.
[4] Abdullah Al-Jafri, personal communication, 2 March 2005.

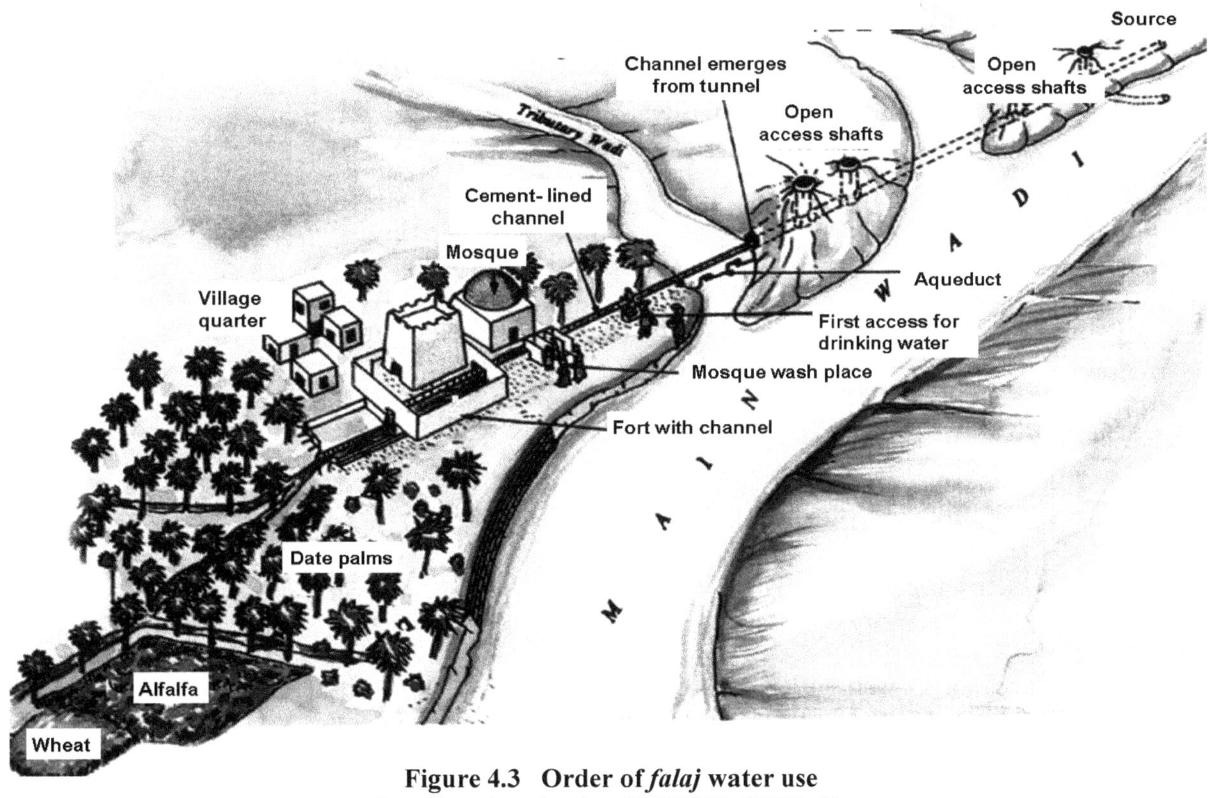

Figure 4.3 Order of *falaj* water use
based on drawing in MRWEWR (2002: 9)

is watered in order, either by position within the irrigated area or according to the ownership of water/time shares. Each garden or individual plot has an offtake from one of the channels and more than one person may be needed: for example, one to change the flow in the main or secondary channels, possibly controlled nowadays by concrete sluices, while another arranges the dams further downstream.

Date palms are the most important crop, not only for the fruit but also for use in many handicrafts such as rope and basket making. Being perennial, date palms require water year round. Seasonal crops such as alfalfa and wheat may be grown downstream of the date palms when there is sufficient water. Some of the water used for irrigation will percolate down past the root zone, and at the downstream end of the cultivated area, excess water either flows to the wadi or provides recharge to the aquifer. Thus *falaj* water may be reused by systems further downstream.

4.4 SIGNIFICANCE OF *FALAJ* WATER TO AGRICULTURE

Since pumps for water wells were introduced to Oman in the 1970s, the use of wells, particularly deeper drilled wells, for agriculture has expanded, resulting in the lowering of groundwater levels and drying up of some *aflāj*. *Falaj* systems have also declined due to many other factors, including the high cost of crop production compared with using wells, and the reducing importance of agriculture as a livelihood. An analysis of these and other factors related to the sustainability of *falaj* agriculture is provided by Dutton (1995) and Al-Shaqsi (1996: 92-111) among others.

Of the c. 1000 "dead" *aflāj* identified in the 1997 National Falaj Inventory Project (NFI), 15% were *ᶜaynī*, 31% *dā'ūdī* and 54% *ghaylī*. This distribution probably reflects the insecurity regarding the flow rate of many *ghaylī aflāj*, and the relative ease of maintenance of *ᶜaynī aflāj*. Despite these problems, *falaj* systems still play a significant part in agricultural productivity in Oman as shown for 1997 in Figure 4.4. Reliable statistics for more recent years are not available, but the situation is unlikely to have changed significantly,[5] even though it is estimated that some 200 *aflāj* have gone out of use since then, mainly due to falling water levels.[6] The *dā'ūdī aflāj* provide more water and irrigate an area greater than that of *ghaylī* and *ᶜaynī aflāj* combined (Al Sulaimani, Helmi and Nash, 2007: 5).

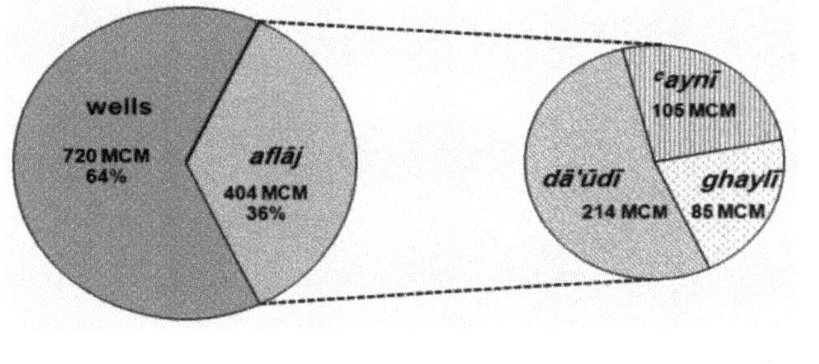

Figure 4.4 Water use in agriculture in Oman in 1997
million cubic metres (MCM) per annum

[5] Hamad Al-Hatmi, personal communication, 25 March 2008.
[6] Tariq Helmi, personal communication, 15 April 2008.

4.5 *FALAJ* ORGANISATION

This section applies to many, but by no means all, of the *falaj* systems of Oman, the main exceptions being those "owned" by a sheikh or by a single family as described by Letts (1978: 317).

The running of the *falaj* system is organised by the community, often by a group of village elders. They agree the appointment of officials responsible for keeping the books, organising repairs and the division of water. Disputes over the allocation of water are normally resolved by the officials, but sometimes the sheikh or even the *walī*, the head of the administrative area, may get involved. Wilkinson (1977: 97-121) describes the system of administration, including the responsibilities of different officials, in some areas of Oman prior to the 1970s, and Letts (1978: 317-321) provides similar information for a different area in the first half of the 1970s, illustrating a wide variation in how different communities administered their system. The fundamentals of the organisation remain the same today, although payment for services is now more often in cash rather than in kind, and many of the wealthier villagers hire people from the Indian sub-continent as farm hands. Many authors use the details given by Wilkinson to describe *falaj* administration, but, as illustrated by Letts, there are many differences from village to village and with time. Information on the administration is therefore given for each study village in the relevant chapter.

Much of the irrigated land and water is owned or rented by villagers, with a portion kept aside for *falaj* maintenance and for payments to the officials. Traditionally, there was an allocation for mosques and their schools and for the poor, but this practice is dying out as people become more dependent on the government. In addition, agricultural produce was taxed prior to 1970 (Wilkinson, 1977: 146-148), and in at least one village some 10% of the crop was given to watchmen to guard the fields[7].

The affairs of the *falaj* are arranged around the irrigation cycle (*dawrān*), the time it takes to water all the land, which typically ranges from about seven to fourteen days (Al-Ghafri, 2004: 47). The higher the flow rate, the more land can be watered and the longer the irrigation cycle. Heavy soils require longer periods of inundation for water to reach the crop roots, so soil type is also a factor determining the length of cycle. The cycle may be varied, the most frequent change being, it seems, to lengthen it during periods of drought, so that the land is irrigated less frequently. When someone refers to their water share as, say, one *athar*, the meaning of which is discussed below, they generally mean one *athar* in each irrigation cycle, and in many places the shares for *falaj* maintenance are auctioned or sold every cycle. The irrigation cycle is also often used as the period for changing the time of day at which a water share is taken, as illustrated in the chapters on the study villages.

Sustainability was the way of life in Oman only 40 years ago: because *falaj* systems use gravity flow, water use never exceeded water supply. The *falaj* system has survived for millennia, through periods of war and destruction, both needing and promoting a cohesive society. Because of their historical and cultural importance, five representative *falaj* systems, together with associated structures such as defensive towers, mosques sundials and buildings for water auctions, have recently been designated as a UNESCO World Heritage site (Al Sulaimani et al., 2007; MRM&WR, 2008).

4.6 TIMING OF WATER SHARES

4.6.1 Divisions of time

Because the water in the *aflāj* flows continuously, it is used at night as well as by day so that it is not wasted. Hence there is a need to distribute the water both day and night, and usually a need to tell the time in order to do so. In many villages, before the advent of watches the sun was used by day and the stars by night to time the allocation of water. The water clock was used in only a few places (Section 3.5.1) and the terms described in this section relate mainly to the use of stars and the sun, rather than the water clock. On cloudy nights, the water clock could be used as usual, but for the stars it was necessary to estimate the time either by using other, more visible, stars or from experience.

The 24-hour day starts either at sunrise or sunset, and is often divided into two parts often called *badda,* one for day and one for night, although in some places the *badda* may be a full day or, at the other extreme, one third of a day (Wilkinson, 1974: 25). Dictionary definitions include أَبَدَّ بَيْنَهُم (*abadda baynahum)* – "to divide among, giving a share" (Badger, 1881: 955), and *baddada* بَدَّدَ (Lane, 1968 1: 161), meaning to divide equally into shares. While the lexica make it clear that the root is √(b.d.d), Wilkinson (eg 1974: 25) uses the form *bāda*, and current usage in both speech and writing in the study villages varies. The form *badda* is used from here on to avoid confusion.

Each day is divided into 48 *athar*s, 24 by day and 24 by night, and so the *athar* is nominally half an hour. People may own shares that are less than one *athar* but $^{1}/_{24}$ of an *athar*, a little more than one minute, is the smallest practicable division, said to be enough to irrigate one palm tree Al-Ghafri (2004: 48), but of course this depends on the flow in the *falaj*.

In the *Encyclopaedia of Islam* New Edition (Index Volume Fascicule 2 Glossary and Index of Terms, 2006: 169), the term *athar* is defined as "a trace; .. a relic of the Prophet ... especially impressions of his footprints". In Wehr it is a track or trace (1966: 4), and in Badger (1881: 1114) "a mark left by something passing", which could

[7] Hamad Al-Hatmi, personal communication regarding his home village - Al Hajar, 17 March 2007.

be taken as describing the path of the shadow of the pole of a sundial, or indeed of any object, and farmers may use their own bodies as a portable sundial (eg Letts, 1978: 322-323).

Subdivisions of the *athar* by different *falaj* communities are described by Wilkinson (1977: 108-109), Al-Shaqsi (1996: 76, 78) and Al-Ghafri (2004: 48) among others, but were not encountered during the fieldwork for this study. The *sahm*, which is not mentioned in the above works, was, however, encountered in several villages. This is a time unit of water equal to 1½ *athar*s, and therefore nominally ¾ hour. In the *Encyclopaedia of Islam* New Edition (Index Volume Fascicule 2 Glossary and Index of Terms, 2006: 482), it is described as denoting a fixed share in law. Badger (1881: 777-778, 955) defines it as a part assigned and as a portion.

4.6.2 Sundials

The sundials in Oman generally comprise a vertical pole and lines on the ground demarking time intervals, which are defined by the movement of the shadow of the tip of the pole. Over the course of a year, the tip of the pole follows different curves: this is clearly seen in the sundial at Qarya after it was freshly painted following resurfacing of the road (Figure 3.5). Qarya is located south of the Tropic of Cancer and so for a part of the summer, the sun is to the north, casting a shadow to the south of the pole. Because the apparent motion of the sun is slower around midday compared to sunset and sunrise, the lines on the ground are close together near the pole and further apart away from the pole to give equal spaces of time. Few sundials have as many divisions as this one; there may only be a few lines but personal markers such as stones (here), coloured plastic or date palm leaves are widely used to show the exact time of someone's water.

The passage of stars and the sun are not the only means of telling the time in the absence of watches. Other methods include the use of water clocks and sand timers. Perhaps surprisingly, sand timers (such as the egg timers used in my childhood) were not invented until the 13th C (Richards, 1999: 52) and are not known in Oman, at least for *falaj* purposes.

4.6.3 Water clocks

The earliest water clock known is thought to be from c. 1500 BC from Karnak (Richards, 1999: 52). This appears to have been of a similar design to that used up to the 20th C in Iran and Oman, and probably elsewhere, for timing irrigation water. Since it has been assumed for a long time that the *falaj* system came to Oman from Iran, one would expect similar methods for timing the distribution of water to be used. However, until recently, it appeared that only water clocks were used in Iran (eg Bonine, 1996: 193). In Oman, the water clock (*ṭāsa*, meaning a shallow drinking cup made of metal) was used in only a few places, including Nizwa (eg Abercrombie, 1981), Saiq and some other villages in the mountains (Al-Ghafri, 2004: 54, 56). It comprised two bowls: one large filled with water, the other small one with a hole in the bottom. When placed in the larger bowl, the smaller one fills with water over a period of time and sinks to the bottom. Al-Ghafri describes the method in more detail. A

Figure 4.5 Sundial at Qarya showing the path traced by the pole's shadow
March 2005

water clock in the Yazd Water Museum is shown in Figure 4.6. This was used for timing irrigation water in Yazd, Iran, possibly in the 1950s according to Majid Labbaf Khaneiki of the International Center on Qanats and Historic Hydraulic Structures, and comprises a small metal cup and larger ceramic bowl. In Nizwa, both bowls were of metal. From Al-Ghafri's work, it appears that the water clock was used in Saiq both day and night, and the time/water shares are also called *ṭāsa*, as shown in a *falaj* book for Falaj al-ᶜAzīzī (Figure 4.7, lines 3, 9 and 11 below the heading). According to Nasr al-Kharusi, *wakīl* (administrator) for Falaj Daris in Nizwa (7 April 2008), the sun and stars were used there before the water clock, which was first introduced probably over 200 years ago for daytime measurements and later for the night.

Figure 4.6
Water clock in Yazd Water Museum

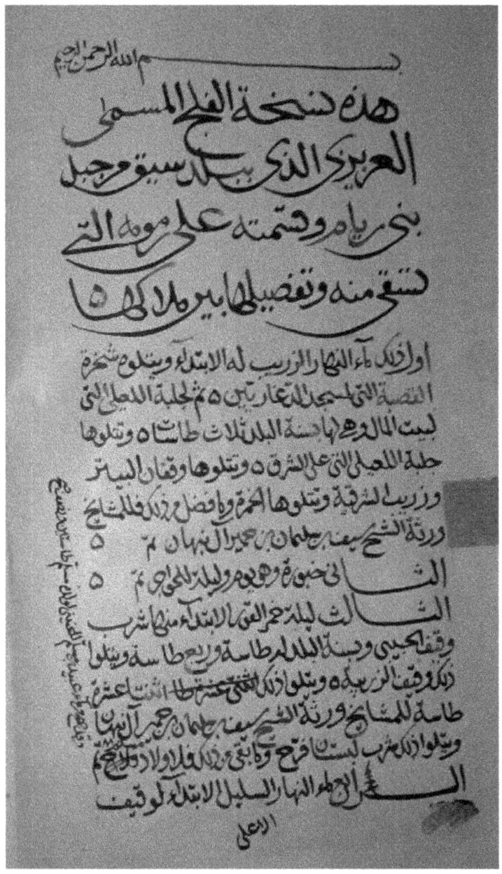

بسم الله الرحمن الرحيم
هذه نسخة الفلج المسمى
العزيزي الذي ببلد سيو وجبل
بني ريام وقسمته على مومه التي
تسقى منه وتفصيلها بين ملاكها

Figure 4.7 (right)
Folio 1 from Falaj ᶜAzīzī *falaj* book
(1308/1891 MS 2316 in Ministry of Heritage and Culture, Muscat)

5 INTRODUCTION TO THE STUDY AREAS

This section gives an introduction to the two study areas, Al Hamra and Mudaybi, and the collection of information there, with some general information on population, *falaj* characteristics and flow, to facilitate a comparison of the villages. The locations of the study areas are shown in Figure 1.3 and the different types of *falaj* are described in Section 4.1.

5.1 AL HAMRA AREA

5.1.1 Background information on the Al Hamra area

Al Hamra is located in the Dakhilyah (Interior) Region. Qarya is the only village in this area known to still use stars for timing water shares. Information was therefore also collected from nearby settlements, Misfat Al-Abriyin (Misfa), Qalat Al-Musaliha (Qala) and Al Hamra itself to obtain a broader understanding of the practice of stargazing in the locality. The locations of these settlements are shown in Figure 5.1.

Qarya, Qala and Al Hamra lie on a plain on the south side and at the foot of the Hajar Mountains, at an elevation of approximately 650 m ASL. Misfa is higher up, within the mountains on the edge of a gorge, at approximately 930 m ASL. Some basic statistics on the settlements and their *aflāj* are given in Table 5.1.

Settlement	Qarya	Misfa	Qala	Al Hamra
Population	1681	615	1334	8453
Falaj name	الكلبي Kalbay	المسفاه Misfa	الحمام Hammam	الحمراء Hamra
Type	*ᶜaynī*	*ᶜaynī*	*dā'ūdī*	*dā'ūdī*
Inventory number	F 0202	F 0241	F 0294	F 0201
Channel length, m^2	790	907	3423	5000
Demand area, ha	12	11	4	10
Irrigation cycle, days	$15^1/_2$	8	11	7

Table 5.1 Basic statistics on Qarya and nearby settlements

The population figures are from the 2003 census; the *falaj* name, type, inventory number, channel length and demand area are from the National Falaj Inventory (NFI; see Section 2.1.2), with the channel length including any branches and taken from the source to the point of first access for domestic use (the *sharīᶜa*, see Section 4.3.2). The length of the irrigation cycle was obtained from interviews.

In Al Hamra and Qala, the relatively small area irrigated by the *aflāj* compared to the population is indicative of a more mixed economy than for Qarya and Misfa, with fewer people dependent on irrigated agriculture. The name al-Hammam for Qala's *falaj* literally means "bath", a term often applied to mineral or thermal waters, and the water has a temperature of 40°C compared with c. 28°C for others in the area[8].

The flow rate of the *aflāj* at Qarya, Misfa, Qala and Al Hamra is monitored by the Ministry of Water Resources (MWR) on a monthly basis, and the hydrographs show a similar pattern in all four *aflāj* (Figure 5.2). This is not surprising: given their proximity, the rainfall over their catchment areas will be similar. The pronounced peakiness of the flow is attributed to the storage characteristics of the aquifers. The most stable flow is at Misfa, indicating that the limestone aquifer supplying the spring used has a greater storage capacity than that supplying the spring at Qarya. The somewhat lesser peakiness of Al Hamra's hydrograph compared to that of Qarya is attributed to the contribution to *falaj* flow from wadi gravels, while the slightly smoother pattern in all, except perhaps Qarya, prior to 1987 is simply a function of the lower frequency of monitoring.

Qala has the lowest flows, and when there is insufficient for irrigation, the *falaj* is supported by water from wells installed by the MWR. The use of hundreds of pumped wells resulted in a reduction of flow in the *falaj* at Al Hamra, and a groundwater recharge dam was built by the MWR in 1989 at Wadi Ghul, upstream of the town. Along with other groundwater recharge dams elsewhere in Oman, it was designed to impound surface water such that the overflow is relatively free of silt and can readily recharge the wadi gravels downstream of the dam. As pointed out by Al-Shaqsi (1996: 53), the flow rate in the *falaj*, which derives its water mainly from the wadi gravels, was generally higher from 1990 to 1996 than from 1982 to 1987 (the flow was not monitored between 1987 and 1990), and he concluded that the recharge dam was effective in increasing the flow of the *falaj*. While this effect may have continued, at least until 2000, the period of monitoring before 1989 is too short to state conclusively that any increase in flow is attributable to the dam. The generally higher base flow seen from 1996 to 1998 in all of the *aflāj* is attributable to higher groundwater recharge in that period, and not especially to the recharge dam, which could have no impact on the flow at Qarya and Misfa.

[8] Hamid Al-Sabarri, MWR Al Hamra, 22 November 2008.

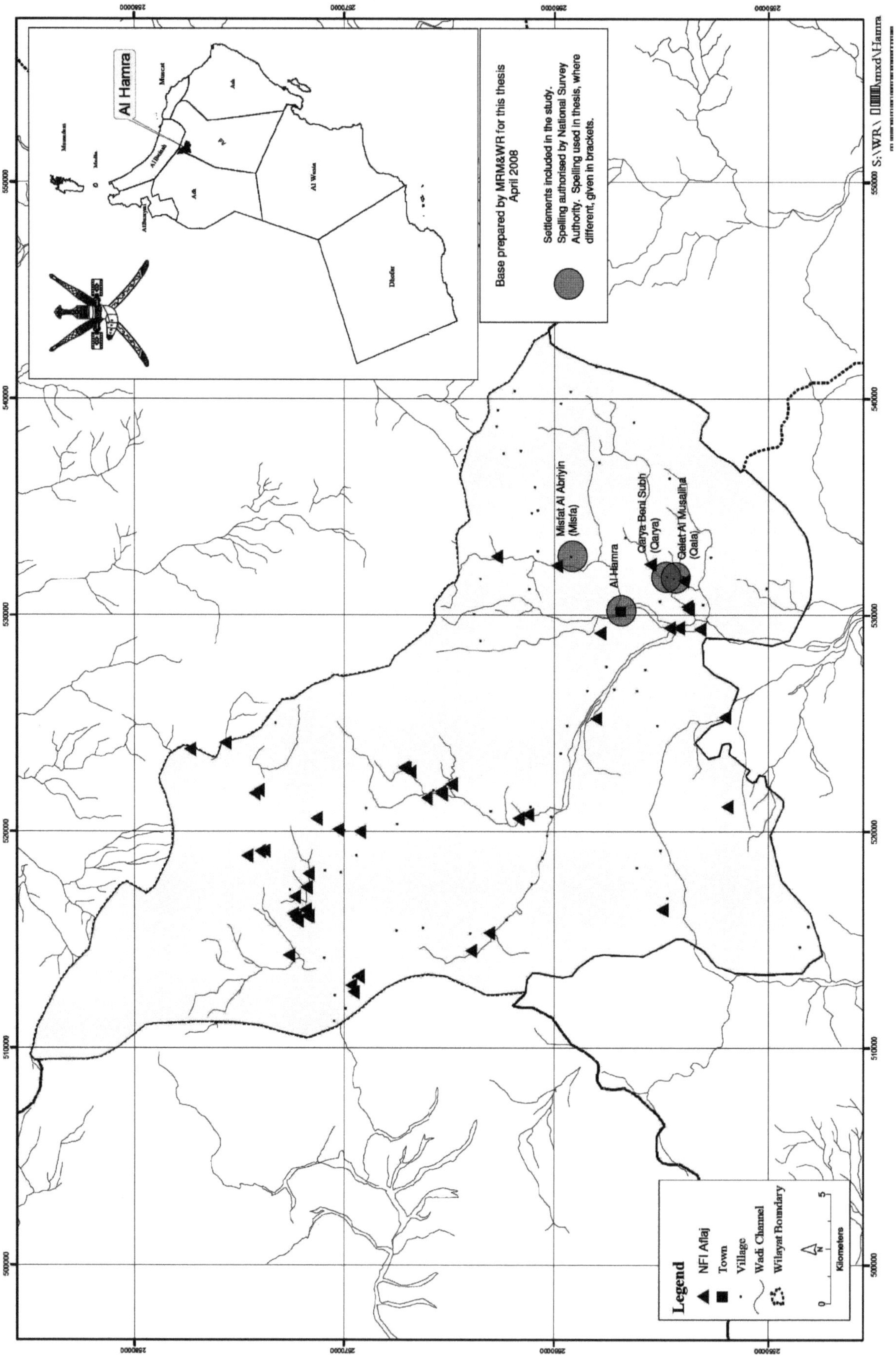

Figure 5.1 **Locations of *aflāj* in the Al Hamra area**

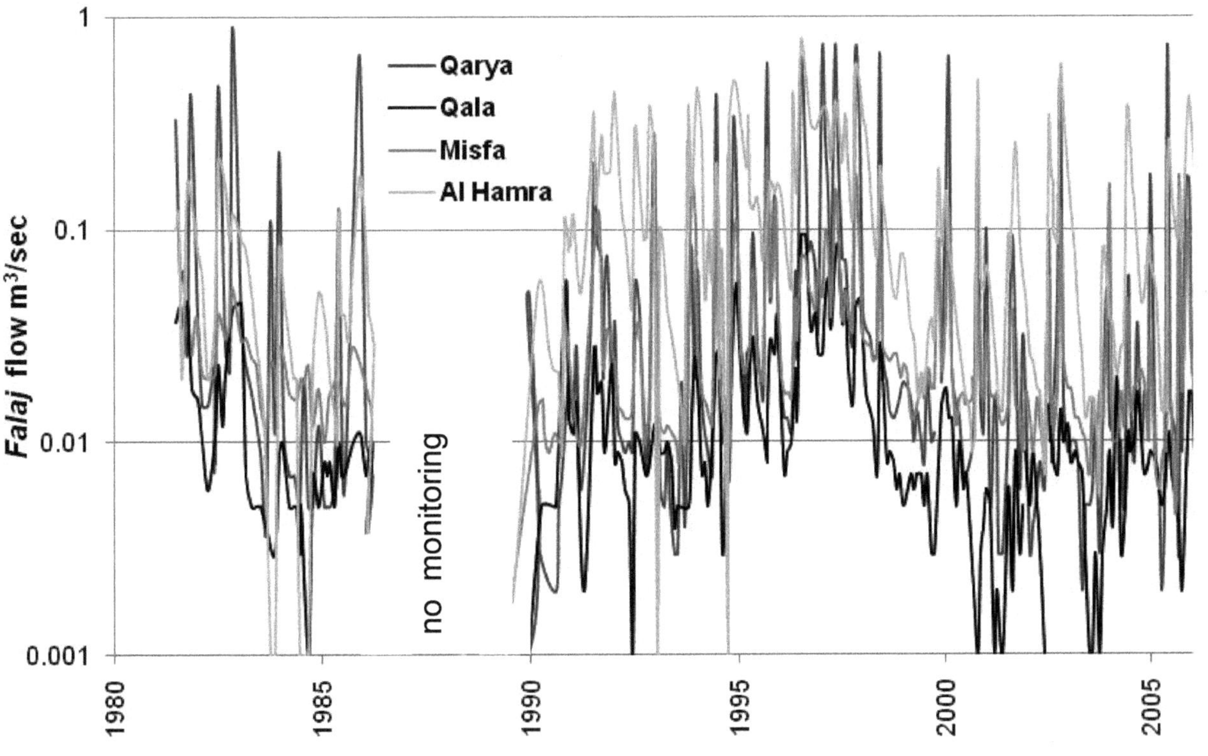

Figure 5.2 ***Falaj*** **flow in the Al Hamra area**
Note: due to the wide range of flow rates, a logarithmic scale is used for the y axis

Falaj flows may also be affected by improvements, particularly extensions to tunnels in the source area or cementing of tunnels in the transport section in *dā'ūdī aflāj*. There is less scope for improvement to *ᶜaynī aflāj*, which depend on natural spring flow, although sealing of channels with cement, as had been done in Qarya, will reduce losses.

5.1.2 Data collection in the Al Hamra area

The names and positions of the main people interviewed, their position related to the *falaj* organization, and their approximate ages in 2007 are given in Table 5.2.

Name	Position/role	Age
Qarya (all of the tribal group Al-Subhi)		
Rashid Khamis bin Huwaishi	*ᶜarīf* responsible for water allocation day and night	55
Saif Salim bin Saif	*wakīl* (administrator)	53
Khamis Huwaishi	Rashid's father, former stargazer	80
Abdullah Salim bin Saif	head of the *falaj* council	57
Misfa		
Masud Salim al-Hattali	knowledgeable about the stars; head of the Hattalis in Misfa	88
Khalfan Salim al-Hattali	Masud's brother; knowledgeable about the stars	66
Qala		
Harith Nahan Al-Abri	*wakīl*	51

Table 5.2 People interviewed in the area of Al Hamra

Initial introductions to the *wakīl* (administrator), Saif bin Salim Al-Subhi in Qarya and village elders in Misfa were made in September 2005 by Ahmed Al-Sabbari, then head of the Water Resource Section in the MRW office in Al Hamra.

In Qarya, by far the most time was spent with the *ᶜarīf* (officer) of the *falaj* and stargazer, Rashid bin Khamis bin Huwaishi, either at the stargazing points or in his home, followed by Saif bin Salim, one of the two *wakīl*s of the *falaj* in his home. Information was collected about the current use of stars and the running of the *falaj*.

In Misfa, Qala and Al Hamra, the information collected focused on the former use of stars. Misfa was allocated the most time of the three because of the interest shown by members of the Historical Association of Oman and others in the towers on the hillside, which were thought to be used as star markers. Several visits were paid to the village, a few to the locations formerly used for watching the stars, but most to the homes of the brothers Masud and Khalfan bin Salim Al-Hattali. Information for Qala was collected from the *wakīl* of the *falaj*, Harith bin Nahan Al-Abri, in his office in the Municipality of Al Hamra, while for Al Hamra, background information on the running of the *falaj* was taken mainly from Al-Shaqsi's (1996) MSc thesis, and information on the stars from documents described below. No attempt was made to repeat Al-Shaqsi's work, which provides a wealth of detail on the hydrology, *falaj* management, water shares and traditional means of timing the shares.

5.2 MUDAYBI AREA

5.2.1 Background information on the Mudaybi area

In the area of Mudaybi, four villages: Zahib, Sudayra, Al Fath and Barzaman, were still using both stars and sundials in 2007. The villages lie on a plain falling from c. 400 m ASL at Al Fath to c. 300 m ASL at Barzaman. The locations of the villages are shown in Figure 5.3 and the basic statistics of their *aflāj* are given in Table 5.3. The population figures are from the 2003 figures provided by the Ministry of Electricity and Water Drinking Water Master Plan 2003-2023 or interviewees; the *falaj* name, type, inventory number, channel length and demand area are from the National Falaj Inventory, and the irrigation cycle from interviewees.

Settlement	Al Fath	Sudayra	Zahib	Barzaman
Population	1749	<1000	<1000	950
Falaj name	الفتح Fath	سدران Sedran	الزاهب Zahib	برزمان Barzaman
Type	*dā'ūdī*	*dā'ūdī*	*dā'ūdī*	*dā'ūdī*
Inventory number	F 2408	F 3475	F 2674	F 1389
Channel length, m²	6291	6355	5499	4801
Demand area, ha	62.2	62.6	26.2	18.7
Irrigation cycle, days	18 or 18½	7	7	9

Table 5.3 Basic statistics for the four *aflāj* near Mudaybi

The flow hydrographs for the *aflāj* are shown in Figure 5.4. The noticeably smoother pattern prior to 1990 is attributable to the lower frequency of monitoring. The period of higher flows in all except Barzaman from c. 1997 to 2002 coincides with relatively high flows in the Al Hamra area, and must be due to higher rainfall throughout the mountains. Zahib and Barzaman generally have lower flows, and therefore smaller areas of cultivation than Al Fath and Sudayra, which show a remarkable similarity in both rate and pattern of flow. This similarity could mean that they are tapping the same aquifer system, possibly the same or connected wadi gravels. The *falaj* at Zahib follows a similar pattern to that at Al Fath and Sudayra but with lower rates of flow. The similarity means that it may be connected to the same groundwater system, and the reduction in peakiness could be because it is further downstream and there may, therefore, be a greater amount of storage in the system. The low flows in all four show a general decline since c. 1990, attributable to a regional decline in water levels due to the use of pumped wells, and much of the peakiness in all of these hydrographs is probably due to the impact of pumping wells in their catchment areas.

5.2.2 Data collection in the Mudaybi area

Initial introductions in Barzaman, Sudayra and Al Fath were made in April 2006 by Ibrahim Al-Namani, water resource technician at the MRW, while he was with me looking for places that still use stars. Later, in November 2006, Ali Al-Habsi from the same office, introduced me to Sultan Al-Shidhani in Al Fath. Introductions to *falaj* officials in Zahib were made in 2005 by the *wakīl* (administrator), Saud bin Abdullah Al-Habsi who, together with the *ᶜarīf* (officer) Khalfan Al-Gabri, provided information on *falaj* management. Here, most of the information on stars and division of water was provided by Saleem bin Rashid Al-Hashimi, a farm worker, and most time was spent with him, mainly at the stargazing point.

In Barzaman, I was introduced to Suleiman bin Saeed bin Saif Al-Aamari, possibly the main stargazer, and later met others with considerable knowledge about the system. Most of the information, both on the running of the *falaj* and stargazing, was collected either at the star-watching location or in the house of Salaam bin Abdullah bin Dhiyab Al-Aamari, the *falaj wakīl* and *ᶜarīf* who lives close by. My visits were mainly at weekends because Salaam works in the Ministry of Religious Affairs in Ibra during the week.

In Sudayra, the *wakīl* of the *falaj*, Hamad bin Shinun Al-Hashmi, a farmer, is knowledgeable about the stars, and was almost my sole source of information. Some information was collected at the sundial and stargazing point, but most meetings were held in Hamad's house.

In Al Fath most time was spent with Sultan Al-Shidhani, a farmer whose late father had been the foremost expert on stars in the town, and who has inherited his father's reputation. Information on the running of the *falaj* was provided by Sultan and by the *wakīl* (administrator) Saud bin Humaid Al-Habsi. Sheikh Sultan bin Mohammed Al-Habsi also helped by clarifying some points of *falaj* history and organisation.

The names and positions of the main people interviewed and their approximate age in 2007 are given in Table 5.4.

Some information was also collected about other settlements in the area of Mudaybi, including Sinaw, Mudaybi itself, and Ridda, one of Mudaybi's satellite villages.

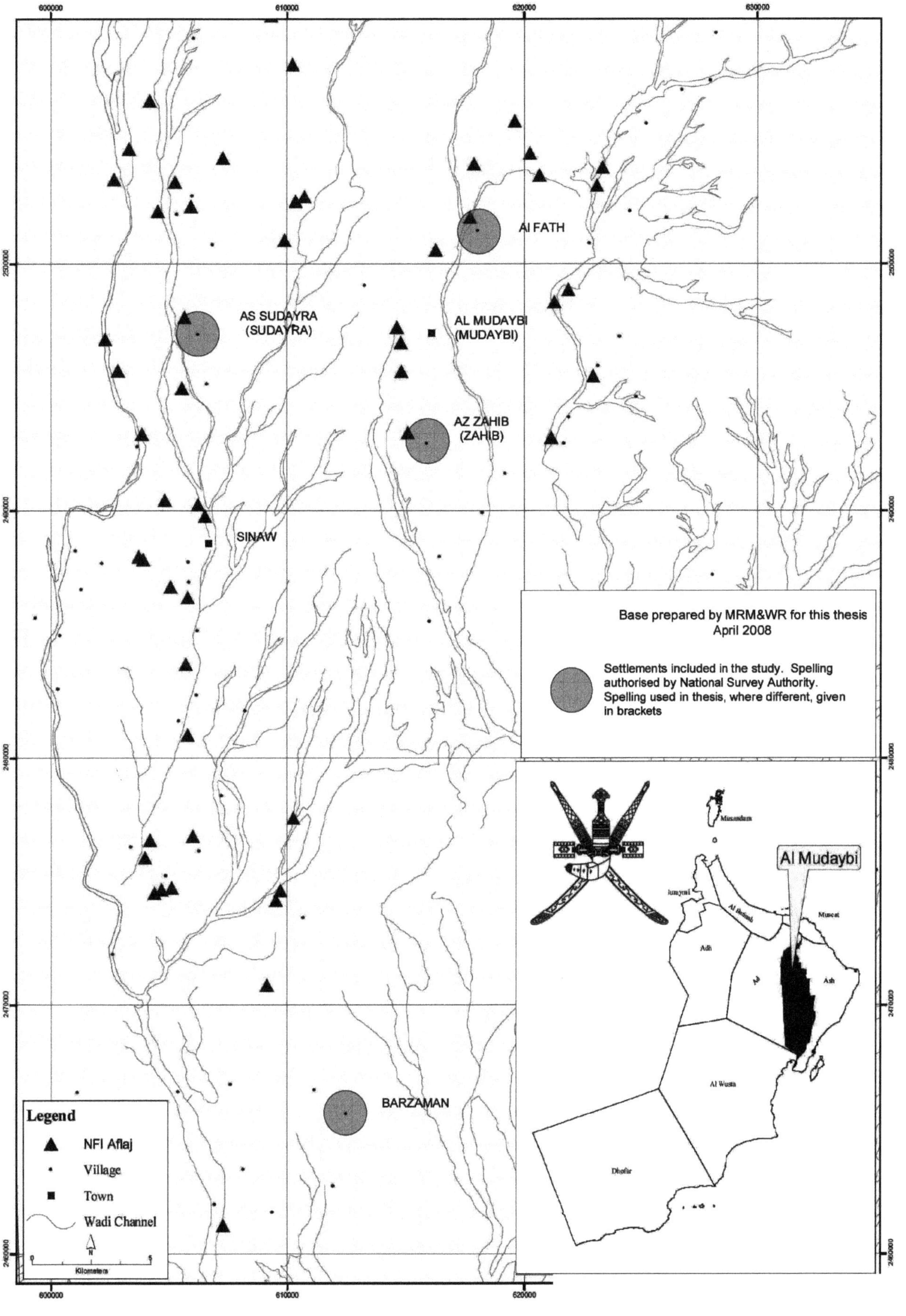

Figure 5.3 Locations of *aflāj* in the Mudaybi area

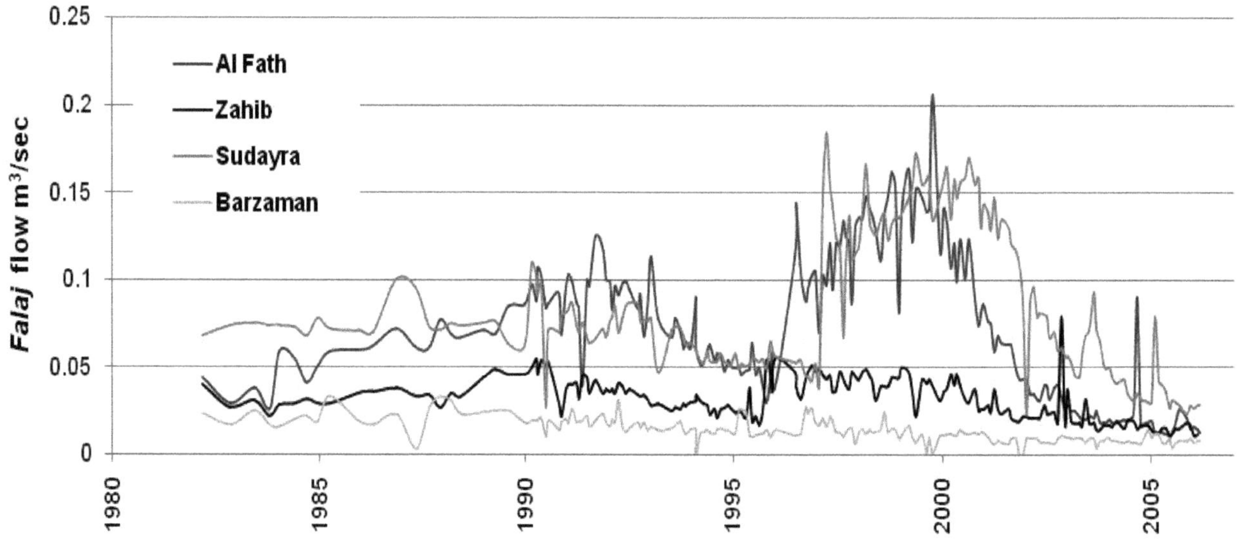

Figure 5.4 ***Falaj*** **flow in the Mudaybi area**

Name	**Position/role**	**Age**
Barzaman (all of the tribal group Al-Aamiri)		
Suleiman Saeed bin Saif	responsible for day and night water distribution; auctioneer	62
Salam Saeed bin Saif	*wakīl* + *ʿarīf* (1 of 2)	50
Al Fath		
Sultan Hamud Al-Shidhani	most knowledgeable about the stars	49
Saud Humaid Al-Habsi	*wakīl*	46
Sudayra		
Hamad Shinun Al-Hashmi	*wakīl* + *ʿarīf* to 2007	60s
Zahib		
Saud Abdullah Al-Habsi	*wakīl*	47
Khalfan Hamud Al-Gabri	*ʿarīf*	65
Saleem Rashid Al-Hashmi	expert on stargazing and water allocation; farm worker	57

Table 5.4 People interviewed in the Mudaybi area

6 AL HAMRA AREA

6.1 INTRODUCTION

The locations of the villages of Qarya, Misfa and Qala, and the town of Al Hamra are shown in Figure 5.1. The inhabitants of Qarya are almost exclusively of the Subhi tribal group. There are two main tribal groups in Misfa: the Hattalis and the Abris. Qala has nine main groups, and Al Hamra is also diverse, but with many Al-Abris. As explained in the previous chapter (Section 5.1), most of the information presented here concerns the village of Qarya.

According to Sheikh Muhsin bin Zahran Al-Abri[9], a respected historian in Al Hamra, and other people interviewed (see Table 5.2), Qarya is an older settlement than Qala or Al Hamra, while Misfa could be as old as, or even older than, Qarya. The *aflāj* at Qarya and Qala are thought to be c. 1000 years old, while that of Al Hamra is more recent: it is said to have been constructed in Yaᶜāriba times when Sulṭān b. Sayf al-Yaᶜrūbī was Imam, from 1060/1650 to 1070/1660 (Al-Shaqsi, 1996: 76). However, Muhsin Al-Abri pointed out that the settlement of Al Hamra was present before, using wells for irrigation. Small scale irrigation from the spring at Misfa may have started a long time ago, but the first development of the *falaj* was in Yaᶜāriba times.

In Qala and Al Hamra, it is thought that the use of the stars and sun for timing the allocation of water goes back as long as the *falaj* systems have been in existence. In Misfa, however, their use did not start until c. 1893 when the system was extended and water shares of less than one day were introduced. Initially, it is possible that the water clock was used in Misfa for daytime measurements and the change to the sundial came later, but stars were always used at night.[10] In Qala, the *ᶜarīf* (*falaj* official often responsible for the distribution of water) taught the farm workers the system, while in Misfa people learned from their fathers or village elders; each person was responsible for their own time keeping.

The approximate year that people stopped using the stars and sun is given in Table 6.1. Al Hamra is the only place among the study villages where watches were introduced for *falaj* timings before the accession of His Majesty Sultan Qaboos in 1970. While filling up my car with petrol in Al Hamra on 4 November 2006, the pump attendant asked about my work. He is from Qarya and said that many people there wanted to change the system to use watches, but that the village elders refused.

The layout of Qarya is shown in Figure 6.1. The village can be divided into three parts: the old village upstream with mud brick houses and now partly in ruins; the newer village with concrete block houses; and thirdly, more recent scattered development in the downstream area. The date palms are conspicuous, with each tree visible, forming a cone downstream of the first main division of the *falaj* channel.

Place	Date use stopped		Comment
	sundial	stars	
Al Hamra	c. 1966	c. 1966	information from Saeed Ali Al-Khiyari, farm worker, 3 April 2007
Misfa	1970s	1970s	
Qala	c. 2000	c. 2000	preceded by a gradual decline in use. No single causal factor

Table 6.1 Use of stars and sundials in settlements near Qarya

The *falaj* is fed by a spring and is called Falaj Al-Kalbay (Chulaybī in the local dialect) meaning "*falaj* of the small dog", which was said to stem from the tradition that a dog found the water source[11], and in November 2006, Rashid bin Khamis Al-Subhi, *ᶜarīf* and official stargazer, showed me the spot where the dog purportedly found and bathed in the water. While most of the land is used for the cultivation of date palms, a significant area is used for seasonal crops, with water from the *falaj* and from wells when there is insufficient water in the *falaj*. Since the source is a spring close to the mountains, there is no significant use of groundwater upgradient of the *falaj*, and the flow has therefore not been reduced by pumped wells.

Unlike settlements with a *dā'ūdī falaj* (a *falaj* collecting groundwater by means of a tunnel), the *sharīᶜa* (i.e. the first main point of access for domestic use by the village, not Islamic Law) on this *ᶜaynī* (spring-fed) *falaj* is downstream of the first point at which the channel is open. The channel is only covered upstream of the first offtake for cultivation, presumably as a protection against debris falling into it. In addition to domestic use within the village, water from the *falaj* is used upstream by pastoralists, who water their animals from openings near the source. Within the village, the *falaj* water is used for many domestic activities, including washing cars and clothes at the *sharīᶜa*, and baths (Figure 6.2). There are several places for washing corpses, but these no longer use water from the *falaj*. Within the irrigated area, the *falaj* channels are widely used for washing.

[9] 8 April 2008.
[10] Khalfan Al-Hattali remembering what his father had told him, 8 April 2008.

[11] Rashid Khamis and Saif Salim Al-Subhi on several occasions. The MSA letter /k/ is often pronounced /ch/ in many places in Oman.

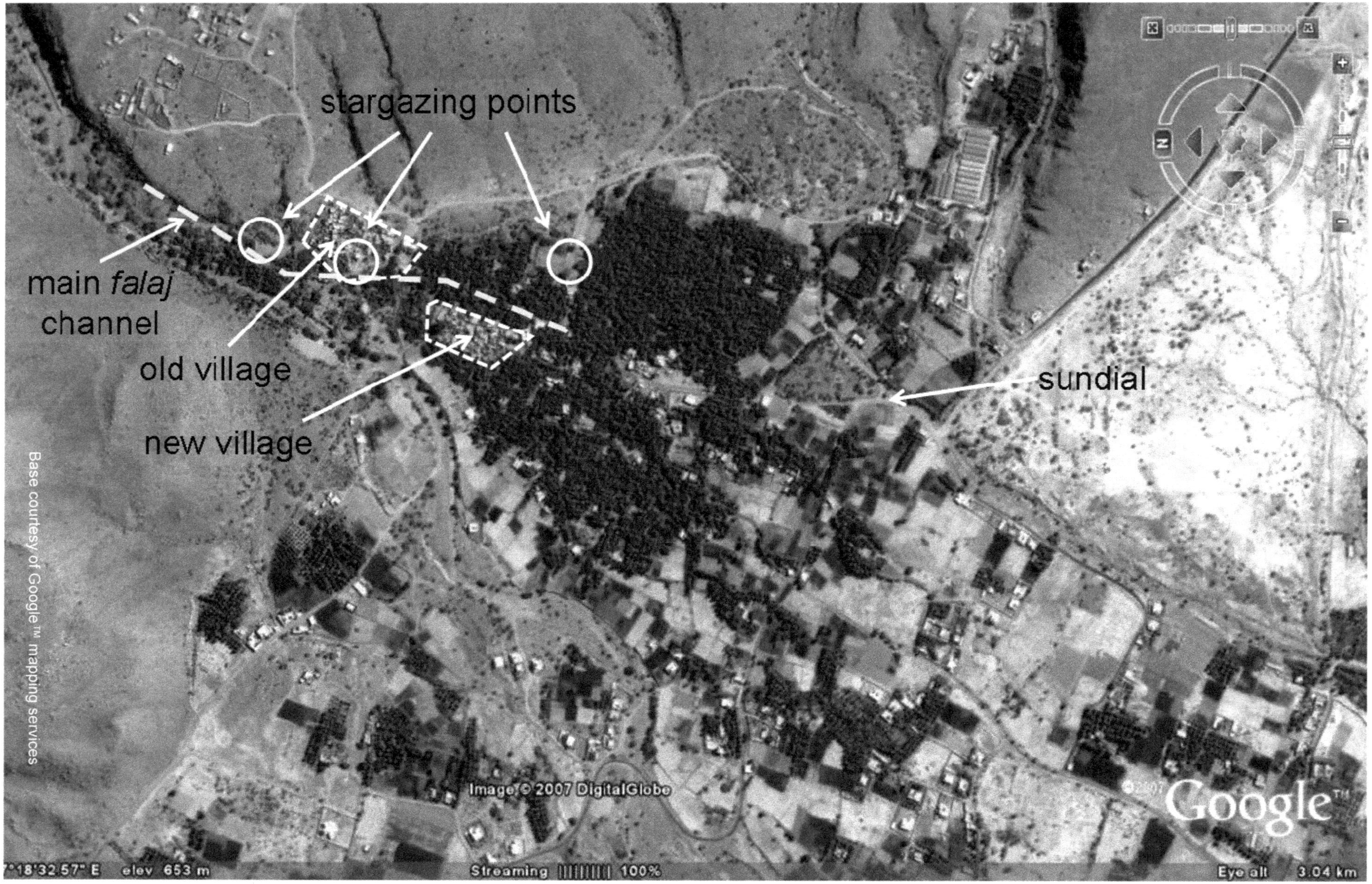

Figure 6.1 Layout of Qarya

Unlike settlements with a *dā'ūdī falaj* (a *falaj* collecting groundwater by means of a tunnel), the *sharīᶜa* (i.e. the first main point of access for domestic use by the village, not Islamic Law) on this *ᶜaynī* (spring-fed) *falaj* is downstream of the first point at which the channel is open. The channel is only covered upstream of the first offtake for cultivation, presumably as a protection against debris falling into it. In addition to domestic use within the village, water from the *falaj* is used upstream by pastoralists, who water their animals from openings near the source. Within the village, the *falaj* water is used for many domestic activities, including washing cars and clothes at the *sharīᶜa*, and baths (Figure 6.2). There are several places for washing corpses, but these no longer use water from the *falaj*. Within the irrigated area, the *falaj* channels are widely used for washing.

6.2 *FALAJ* ORGANISATION IN THE AL HAMRA AREA

The organisation of the *aflāj* varies among the settlements, but they each have a *wakīl* (administrator). Exceptionally, there is no *ᶜarīf* (officer) for Misfa, the village being too small to need one. Information on the main officials responsible for *falaj* management in Qarya is given in Table 6.2. Saif bin Salim is the *wakīl* for the upper part of the village while Zahran bin Khalif is *wakīl* for the lower part. Before Zahran started, Saif was the only *wakīl*. Rashid's father, Khamis, was the previous stargazer but was not an *ᶜarīf*: these posts are not hereditary but are agreed by the *falaj* council, which comprises up to 40 village elders, chief of whom is Abdullah bin Salim, Saif's brother.

In Qarya, the irrigation cycle (*dawra/dawrān*, meaning a cycle or revolution) is 15½ days divided into ½-day units, from sunrise to sunset and from sunset to sunrise, called *bada*s or *bādda*s (i.e. *badda*, see pages 24-25), which here are not given individual names. Initially, it was said that the cycle alternates between 15 and 16 days, but it appears that this description was given to make it easier for me to understand. The term *radd* is also used, and is said to mean the same as *badda*.

Each *badda* is divided into 24 *athar*s. A *sahm* is 1½ *athar*s, but seems to be used less frequently than the *athar*, and by day rather than by night: the main water shares (see Tables 6.3 and 6.4 below), for example, were given in *athar*s. The odd number (31) of *badda*s means that the water shares automatically alternate between day and night with each successive irrigation cycle, and the water is taken from the same *athar*(s) after sunrise or sunset (Figure 6.3).

Figure 6.3 (above) **Rotation of water shares in Qarya**

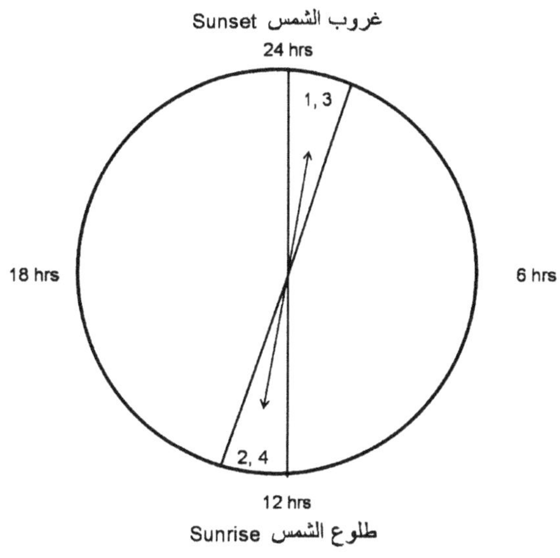

Figure 6.2 (left) **Water use on Falaj al-Kalbay**
Above: point of water use by pastoralists, 10 November 2006.
Below: the car wash (*sharīᶜa*), 11 November 2006

Title	Name (all Al-Subhi)	Role	Year born	Started	Other work
wakīl	Saif bin Salim bin Saif	overall administration including accounts	c. 1955	c. 1997	driver of school bus + farmer
wakīl	Zahran bin Khalif bin Aamar		c. 1955	c. 2000	farmer
ʿarīf	Rashid bin Khamis bin Huwaishi	distribution of water, day + night	c. 1952	c. 1985	cleaning *falaj* channels and timing water
		stargazer			
dillāl	Masud bin Marhun bin Masud	caller of lots of water for auction each irrigation cycle	c. 1937	c. 1992	cleaner for the Municipality

Table 6.2 Officials concerned with *falaj* organisation in Qarya 2007

In Misfa, the water shares also alternate between day and night each irrigation cycle, which is currently eight days. However, there was no division of water rights into *athar*s until 113 or 114 years ago (c. 1893), a time confirmed on separate occasions, and also calculated by Masud Al-Hattali (on 9 May 2006) from his memory of generations of ancestors. Before this, each area of land had water for a day at a time. Then the area of irrigation was extended and it became necessary to allocate water shares at shorter time intervals.

The system for alternating shares in Qarya is simple compared with that in Al Hamra (Al-Shaqsi, 1996: 81), where, for five days of the seven-day irrigation cycle, a person's share alternates between day and night and different quarters of the day or night for successive cycles; however, for two days, the share does not alternate between day and night, although it is taken from different quarters. Taking water at different times of the day means that over a period of one year the average length of an *athar* share of water will be close to ½ hour.

Except when it is cloudy, everyone (i.e. those using *falaj* water for cultivation) in Qarya uses the sundial. It is located near the downstream limit of *falaj* cultivation (Figure 6.1), and is deservedly a source of pride to the village and often commented on by people elsewhere. It is in the middle of a road junction (Figures 3.5 and 6.4), with a tree on the right that provides shade, and consists of a steel pole 1.64 m high and painted lines on the ground. The distance between three red lines marks one *sahm* (c. 45 minutes). Once the shadow reaches the awaited point, the watcher walks at a brisk pace to the village, and for lower land drops dry leaves and twigs (*waraqa*) in the *falaj* channel; when these markers reach the appointed division, the dam is moved. For intermediate land, the dams are changed directly, without the use of *waraqa*. The mark watched for at the sundial takes into account the time for the above procedures, i.e. it is before the start time of that person's water. When it is cloudy, people now use wristwatches.

Left: the pole of the sundial, 7 November 2006 pm
Right: Yasir bin Ali Al-Subhi waiting his turn, 9 May 2006 am

Figure 6.4 The sundial at Qarya

Two different sets of the names of owners of water shares Qarya were given verbally by Rashid bin Khamis and Khamis bin Huwaishi (on 9 November 2006). The first (Table 6.3) is said to be perhaps 1000 years old, in which case it represents the early division of the *falaj* water. The people named are living, and it is inferred that they are the direct descendants of the original shareholders. The first *athar* starts at sunrise, and there are 48 *athar*s in the day: 24 for daytime and 24 for the night. The allocation is said to be the same every day, presumably excluding any day(s) set aside for *falaj* maintenance.

Athar	Owner	*Athar*	Owner
sunrise to sunset		sunset to sunrise	
1	Khamis bin Huwaishi	25	Ali bin Abdullah
2	Mohammed bin Saeed	26	Sheikhan bin Suleiyam
3	Khalif bin Suleiman	27	Khalif bin Khamis
4	Khalid bin Saif	28	Hilal bin Salim
5	Ali bin Hamad	29	Rashid bin Mohammed bin Saeed
6	Salim bin Juma	30	Hamad bin Sleem
7	Abdullah bin Ayisa	31	Saeed bin Abdullah bin Adeen
8	Hamad bin Salim	32	Abbas bin Mansur
9	Khalif bin Salim bin Saeed	33	Suleiman bin Salim
10	Khamis bin Mohammed	34	Aamar bin Salim
11	Azzab bin Khamis	35	Alei bin Salim
12	Abdullah bin Salim bin Saif	36	Saeed bin Suleiman
13	Mohammed bin Ziyad	37	Saif bin Salim
14	Ali bin Hamud	38	Salim bin Hamad
15	Saeed bin Salim	39	Salim bin Nasr
16	Khalif bin Salim bin Rashid	40	Khaamis bin Amr
17	Khalif bin Zahar	41	Hilab bin Mohammed
18	Saeed bin Imbarich	42	Yaqub bin Ali
19	Suleiman bin Zahar	43	Yusuf bin Saeed
20	Saif bin Saleh	44	Yunis bin Mattar
21	Khalif bin Aamar	45	Salim bin Suaid
22	Zahran bin Khalif	46	Abdullah bin Salim bin Nasr
23	Abdullah bin Suleiman	47	Mohammed bin Nasr bin Abdullah
24	Mohammed bin Abdullah	48	Ali bin Suleiman

Table 6.3 Early *falaj* water ownership in Qarya

Current (2006) ownership is different, as shown in Table 6.4, which lists people owning two *athar*s or more in each irrigation cycle. Among these major shareholders are two women: women work in the fields and come to Rashid bin Khamis for clarification of the time for their water as needed, but use an agent to buy water at auction. The *waqf* property (property of the mosques for their own running and for charitable purposes) comprises thirty *athar*s for the mosques and thirty for the poor. This is separate from the 48 *athar*s used for *falaj* maintenance. The total, including the share for *waqf* and *falaj* maintenance, is 502 out of a total of 744 *athar*s for the full irrigation cycle (48 *athar*s/day x 15½ days = 744), from which it can be deduced that most people own less than two *athar*s.

No. of *athars*	Name	No. of *athars*	Name
30	Khamis bin Huwaishi	6	Mohammed bin Ziya
30	Khalif bin Suleiman bin Mansur	6	Saeed bin Salim
30	Khalif bin Zahar + brother	6	Abdullah bin Aisa
c. 25	Ali bin Salim bin Saeed	6	Ali bin Hamad
20	Ali bin Hamid	6	Salim bin Juma
20	Awlād (sons of) Ghanim	6	Khalif bin Khamis
c. 20	Abdullah bin Salim + Salim bin Nasr	6	Awlād Muhay
20	Khalif bin Salim	6	Abdullah bin Suleiman
20	Abbas bin Mansur	5	Salim bin Sabi
15	Khalif bin Aamar	4½	Nauma bint Mohammed
15	Saeed bin Mubarich/ Imbarich	4½	Butay bin Humaid
15	Saeed bin Abdullah bin Adin	4	Abdullah bin Salim
10	Saeed bin Abdullah bin Ali	4	Awlād Saleh bin Mansur
10	Ahmed bin Saeed	3	Suleiman bin Salim
10	Mohammed bin Saeed bin Hamud	3	Khalif bin Saif
10	Akhū (brother of) Yusuf bin Saeed	3	Ali bin Suleiman
7	Khamis bin Salim	3	Aatib bin Salim
7	Khalif bin Salim bin Rashid	3	Saif bin Sahluf
7	Hilal bin Salim bin Rashid	2	Awlād Mayuf
7	Awlād Nasr bin Salim	2	Braich bin Suleiman
		2	Taliba bint Ali

Table 6.4 Main owners of *falaj* water rights in Qarya in 2006

The times from dawn until a shadow can be seen at the sundial, and from the last shadow to the first star are each treated as one *athar*, the first lasting approximately one hour and the latter 40 minutes. Because of the method of alternating water allocations between day and night, the same people will benefit from these longer *athar*s, and they therefore cost more to buy outright, at c. 2000 riyals compared to c. 1000 riyals for shorter ones (Rashid Khamis and Khamis Huwaishi Al-Subhi, 31 October 2007).

There is no *falaj* book detailing water ownership for Qarya, but transactions are ratified by the legal department in the Ministry of Justice. Figure 6.5 gives the text of such a letter written on Ministry of Justice headed paper.

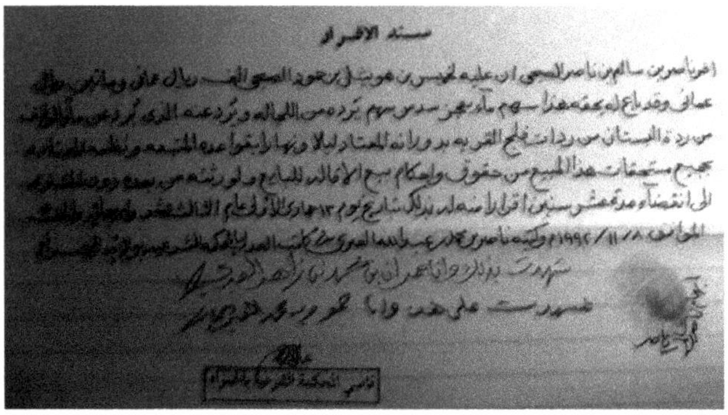

Figure 6.5 Document for sale of water in Qarya
from Khamis bin Huwaishi Al-Subhi, 31 October 2007

In summary, it states that:

> *Nāṣr bin Sālim bin Nāṣr al-Ṣubḥī confirms that he has received from Khamīs bin Huwayshī bin Ḥamūd al-Ṣubḥī 1200 Omani Riyals, and he has sold him the ownership of this* sahm *of water less one sixth of a* sahm *that he retains from ? the circulation... and he retains the* waqf *water from the garden* radda *from the* raddas *of Falaj Qarya ... ... 10 years on 13 Jumād ... 8/11/1996 AD, and this was written by Nāṣr bin Muḥammad bin ᶜAbdallāh al-ᶜAbrī ...*

This is interesting for its use of the term *sahm* rather than *athar* for the water shares, and *radda* (pl. *raddāt*), which here refers to a particular area irrigated by the *falaj* rather than a time share (Rashid Khamis and Majid Khamis Al-Subhi, 5 April 2008).

In Qarya, one day (48 *athar*s) is auctioned for use by the *falaj* each irrigation cycle (Figure 6.6). Only men attend the auction, but women can participate by asking someone to bid for them. People meet under a mango tree outside the village hall, and the auctioneer, Masud bin Marhun, wearing a red lungi in the photograph, calls for bids for 16 lots of three *athar*s each starting at sunrise. The winning bid is recorded separately by the two *wakīl*s. Afterwards, they and the auctioneer check the information and cash received using old quarter annas, currency up to the 1950s, as a tally, each coin representing three *athar*s.

Figure 6.6 Auction of *falaj* water in Qarya, May 2006
Middle, sitting facing tree: Saif Salim, *wakīl*; Right, sitting Zahran Khalif *wakīl*

For the twelve months from January 2004 to January 2005, 1684 riyals were raised at auction, but the amount obtained on any one auction day varied considerably, from 13 to 158 riyals. On 12 May 2006, the amount paid for each lot of three *athar*s ranged from 5.2 to 7.0 riyals. Rashid Khamis was not sure of the reason for the variation in price but speculated that the lower prices were for water further down the system, which would have more soap in it. The main outgoing is for cleaning silt from the channel, an activity called *shiḥāba* in speech, *shuḥūba* in writing, meaning pale or wan (eg Wehr, 1966: 457), possibly related to the colour of the silt, and other maintenance. A summary of *falaj* finances is given in Table 6.5.

Income riyals	Source	Expenditure riyals	Purpose
1680	auctions (*qaᶜada*)	c. 350	payment to officials for auctions and keeping records
		c. 1000	maintenance: cleaning silt and cementing channels
		variable	10% of cost of major works (to the *wakīl*s for supervision etc.)

Table 6.5 Summary of *falaj* finances in Qarya for 2004

The *wakīl*s record the income from water auctions and outgoings, each in their own book. A page from a book kept by Saif Salim is shown in Figure 6.7. On the right is the amount that shares were auctioned for on 23 January 2004, totalling 28.30 riyals, and on the left, at the top the amount received from the auctioneer of 24 riyals, and on the bottom left, a sum of 10 riyals paid for channel maintenance and cleaning. The difference between the amount received at auction and that handed over by the auctioneer probably represents the payment to the auctioneer and *wakīl*s for their services.

In Al Hamra the *falaj* share is a little more than half a day each cycle (Al-Shaqsi, 1996: 79). In Misfa, funds for *falaj* maintenance are obtained from the auction of 16 *athar*s every four to five years.[41] In Qala, 24 hours is auctioned every six months, which raises 1000 riyals each year.[42] This is sufficient for cleaning and maintenance but not always for major works. In the 1980s, the Ministry of Agriculture funded major works in Qala: one or more tributaries were lengthened and distribution channels in fields were repaired. This implies that the *falaj* did not supply sufficient water. The *wakīl* is not paid, but the *ᶜarīf* receives 4% of any profit accumulated during the year, unlike Qarya where the *wakīl*s are paid a small amount and the *ᶜarīf* nothing.

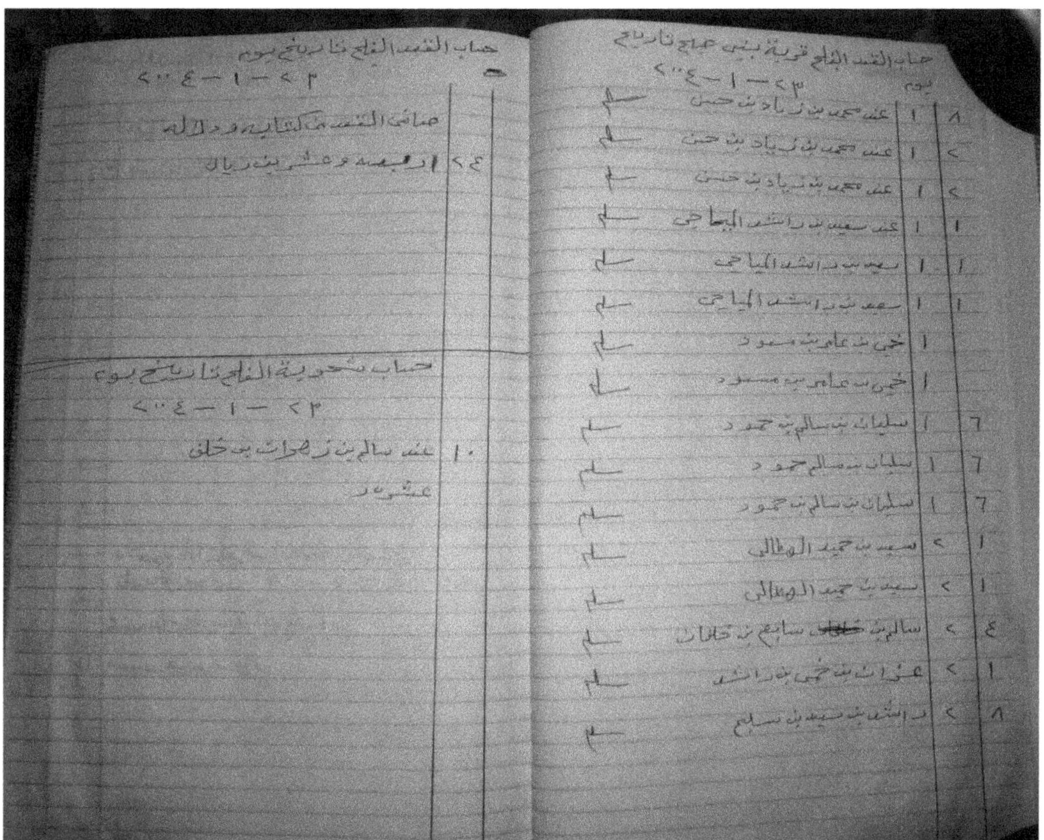

Figure 6.7 Qarya *falaj* accounts book
from Saif bin Salim Al-Subhi, 3 April 2007

[41] Masud and Khalfan Al-Hattali, 30 October 2007.
[42] Harith bin Nabhan Al-Abri, 28 October 2007.

6.3 METHOD OF STAR USE IN THE AL HAMRA AREA

In Qarya, the stars are watched rising in the east above the horizon, and the time between the rise of different stars is used for the division of water. There are three "official" points for watching the stars, the locations of which are shown in Figure 6.1 above. The upstream point is the courtyard of an old mosque, the first building in the village as one walks down the *falaj* channel; the intermediate point is on the roof of the *majlis* (here equivalent to a village hall) for the upper village, and the lowest is outside a house in the newer part of the village. At all three places there are few visual obstructions between the observer and the horizon.

The different viewing points are used at different stages in the irrigation cycle, to be close to the area being irrigated at the time. The upper point is only used for a few days and the intermediate point for little more. Most of the time, the lower point (Figure 6.8) is used. The telegraph pole is not a marker, but provides a useful point of reference for people such as myself learning the system. The main feature to note is that the land rises to the north, towards the mountains and northern stars will therefore rise later than if the horizon were flat. On cloudy nights more visible stars may be used, or more frequently these days, wristwatches.

Rashid bin Khamis bin Huwaishi Al-Subhi (Figure 6.9), elsewhere referred to as Rashid Khamis or Rashid bin Khamis, is the official stargazer in Qarya. He was born in c. 1952 and took over from his father as stargazer in c. 1985. However, the stargazer before this was from a different family. Rashid stays out all night, sometimes taking a nap between stars, sometimes cleaning the irrigation channels, so he is not always at the official stargazing point. Farmers come to him for information as needed: it appears that most people know enough about timing by stars to carry out their own distribution of

Figure 6.8 Horizon used for stargazing at Qarya looking east (January 2006)

water and use Rashid mainly to check when they are unsure. At night there is no use of *waraqa* (dry twigs), people going directly to the appropriate sluice to change the flow of water in the *falaj*.

Figure 6.9 Rashid bin Khamis Al-Subhi, stargazer in Qarya (January 2006)

Rashid Khamis has not found anyone willing to succeed him: although he is teaching his brother and son the system, both say that they would not want to carry on his work as it is unpaid and too hard, especially having to be available all night.

In Misfa, Qala and Al Hamra, the stars were also watched rising above the eastern horizon. In Misfa, four towers were built on the ridge forming the horizon to assist people with poor eyesight. The stars marked are Chuwī (α Lyrae), Thurayyā (the Pleiades), Shubcha (λ, φ^1 + φ^2 Orionis), Mīyazīn (Orion's Belt: δ, ε + ζ Orionis) and ᶜAqrab (β, δ + π Scorpii), as shown in Figure 6.10. The tower for Mīyazīn (meaning weighing scales, but no relation to the constellation Libra), which was used as a divider, has a hole through which one can see the sky, but the hole was not used for watching stars. There are a few

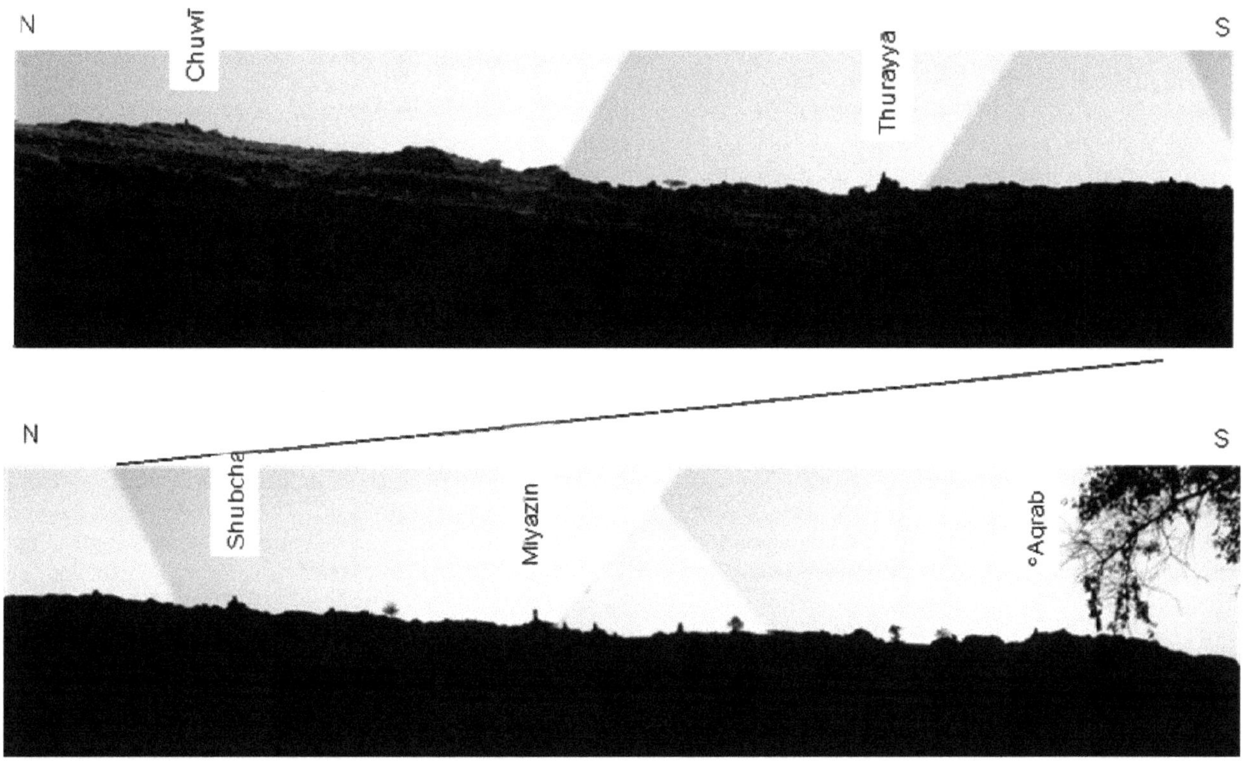

Figure 6.10 Horizon and towers used for watching stars in Misfa

other towers that appear not to be used – it was thought by Masud and Khalfan Al-Hattali, when interviewed on 9 May and 29 November 2006, that they could have been built by children to copy the others.

The mosque was the main stargazing point in Misfa, but it appears that people also watched from their fields and would use whichever star was visible. When the main star could not be seen, others were used, including stars in the Plough and in the zenith. It is often clearer looking at the zenith than at a horizon, especially if the horizon is distant, as the atmosphere is thinner and there is less scattering of light.

According to Saeed bin Ali Al-Khiyari, a farm worker in Al Hamra, Misfa took the system of telling the time with stars from Al Hamra, but neither Masud nor Khalfan Al-Hattali agreed with this.

The use of the term zenith merits further attention. The zenith of the heavens is the point in the sky immediately above the observer. In MSA the term for the astronomical zenith is *kibd al-samā'*, *kibd* meaning both liver (the body organ) and middle/centre/zenith and *al-samā'* the sky or heavens. This term was used both in interviews in Misfa and in written works such as a manuscript from Al Hamra described in Section 6.4.3 below, where it actually means the highest point in the sky reached by a star, i.e. its culmination rather than the zenith, which the star may be nowhere near. To simplify matters the term zenith is used throughout this text, as in Oman, to mean both the astronomical zenith and the culmination of a star.

6.4 STAR NAMES, IDENTIFICATION AND TIMINGS IN THE AL HAMRA AREA

This section presents the information on the stars used, their identification and timings. It starts with Qarya followed by Misfa, in both of which stars were identified during the field work for this study. These are followed by Al Hamra, for which a modern star schedule and older manuscript description of the stars are available, and then by Qala, for which only a modern star schedule was obtained.

The meaning of the star names is not always clear: most people in Oman using stars do not think about their meaning, and for many there is no clear definition given in dictionaries or lexica. This is not dissimilar in the English speaking world: while the meanings of constellation names, such as Cancer (the Crab), are widely known, those for individual star names are not. In these sections on the study areas, meanings of star names are, therefore, given only sparingly, but they are addressed in more detail in Chapter 11.

6.4.1 Stars used in Qarya

Twenty-one main stars are used in Qarya throughout the year to divide the allocation of water. In November 2006, Rashid Khamis's brother, Majid, gave me the star list and schedule for the village, in Arabic, shown in Appendix A.1. The list was said to have been written in 1989. It gives the star name, the number of stars that the name refers to, the start and end date for the star, and the number of days in that period, followed by the number of *athar*s between one star and the next. Notably, the start date of each star is the day on which it is the last star to be seen rising at dawn, i.e. its heliacal rising, and not when it is the first star to rise after sunset. Most of this information is given in Table 6.6, together with the time between the stars given verbally by Rashid Khamis on different occasions. The star names are sometimes different in speech than in writing, for example /ch/ often replaces the MSA sound /k/. However, even the written versions of the star names are not all found in MSA: the MSA for Shubka, for example is Shābik (meaning "fishing net").

The times between the main stars range from 40 to 115 minutes, and differ slightly from those given for Qarya by Al-Shaqsi (1996: 85), who obtained his information from Rashid Khamis's father, Khamis bin Huwaishi. The time between the rise of Shaᶜarā (Procyon, α Canis Minoris) and Janb (δ, ε + ρ Hydrae) is readily checked since they rise at the same latitude and the shape of the horizon has little effect on the relative time of rise. I could not see the rise of the these stars on 12 January 2006 but Starry Night™ shows that there are 58 to 68 minutes between them, indicating that the information given by Rashid Khamis is reliable.

There is some discrepancy between the dates given in the schedule and the number of days until the next star becomes the first to rise at dawn for Kawkabayn, Sufāᶜ and ᶜAqrab, which Rashid Khamis attributes to an error in the dates, not in the number of days. The total number of days is 368, which seems to be due to rounding rather than an error in the account.

Table 6.7 gives the names of the stars used in Qarya, their identifications established during this study, and the time that they represent. The magnitude (brightness) is also given, since it could be pertinent to the selection of stars: the faintest star used on its own, i.e. with no brighter star to help, is 1 Pegasi, the divider between Mūfī and Ghurāb, with a magnitude of 4.06. The time between the main stars is split into *athar*s lasting from 20 to 50 minutes, often using divider stars, which are not given names; they are known only as the divider (*munṣaf*, pl. *manāṣif*) of the previous star. Additional stars are used as part of the time telling process but are not *athar* dividers, and not all *athar*s are divided by stars, so there is some scope for confusion. The times between stars in Qarya in Tables 6.6 and 6.7 are given in minutes, but time in the field is reckoned in *athar*s, which can be confusing since the *athar*s are of different lengths. For example, the two stars of Kawkabayn are said to be ¼ plus ⅛ of an *athar* apart, which for a 25 minute *athar* is just over 7 minutes, and for a 30 minute *athar* is nearly 11 minutes. One might expect the variation in *athar* length to reflect the length of the night in different seasons, but it does not. However, since different stars are used at different times of the year, the *athar* length averages out to be roughly equal over the course of one year.

Star name (written)		**Star name**** (phonetic representation)	**Start*** (d/m)	**Number of***			**Time to next star min** ***
Arabic*	Transliteration			**stars**	**days**	***athars***	
كوي	Kuwī	Chuwī	21/12	1	24	3	115
الموفي	Mūfī	Mūfī	14/1	1	16	2	40
الغراب	Ghurāb	Ghurāb	30/1	1	16	2	75
الادم	Adam	Adam	15/2	1	16	2	45
الصارة 1	Ṣāra 1	Ṣāra (*al-awliyya*)	3/3	1	16	2	65
الصارة 2	Ṣāra 2	Ṣāra (*al-akhīra*)	19/3	1	16	2	40
السعد	Saᶜd	Saᶜd	4/4	1	16	2	70
الكوكبين	Kawkabayn	Chūchabayn	20/4	2	24	3	90
الثريا	Thurayyā	Thurayyā	16/5	9	16	2	60
الدبران	Dabarān	Debrān	1/6	1	16	2	70
الشبكة	Shubka	Shubcha	17/6	3	16	2	50
الجوزاء	Jawzā'	Jawza	3/7	1	16	2	75
الشعرى	Shaᶜarā	Shaᶜara	19/7	1	16	2	60 (80)
الجنب	Janb	Janb	4/8	1	16	2	70 (60)
الذراع	Dharāᶜ	Dharāᶜīn	20/8	2	16	2	55
البطين	Buṭayn	Buṭayn	5 /9	1	16	2	40
السفاع	Sufāᶜ	Sufāᶜ/Sifāᶜ	21/9	1	16	2	80
المواثيب	Mawāthīb	Mīyathāb	6/10	7	16	2	60
الذكرين	Dhakarayn	Dhakarayn	22/10	2	24	3	70 +/- 2
الغفر	Ghafr	Ghafur	15/11	1	16	2	90
العقرب	ᶜAqrab	ᶜAqrab	1/12	2	24	3	70

* from star schedule (Appendix A.1); ** as spoken by Rashid bin Khamis Al-Subhi; *** information from Rashid bin Khamis; times given by Al-Shaqsi (1996: 85) in brackets where 10 minutes or more difference.

Table 6.6 Main *falaj* stars identified in Qarya

Star			Magnitude	*Athar* length, min			Time to next star, min
Name (from Table 6.6)	**Classification**	**English name**					
Chuwī	α Lyrae	Vega	0.00	40	40	35	20
divider	β Lyrae	Sheliak	3.50				10
divider	γ Lyrae	Sulafat	3.25				30
divider	δ Cygni		2.84				30
divider	o^1 Cygni		3.78				15
	γ Cygni	Sadr	2.21				10
Mūfī	α Cygni	Deneb	1.25	20	20		20
divider	ε Cygni	Gienah	2.46				20
Ghurāb	α, β, δ + λ^2 Delphini	Sualocin (α), Rotanev (β)	3.62 - 4.40	40	35		40
divider	1 Pegasi		4.06				35
Adam	ε Pegasi	Enif	2.37	20	25		20
divider	η Pegasi	Matar	2.90				25
Ṣāra (*al-awliyya*)*	β Cassiopeiae	Caph	2.25	32.5	32.5		25
	α Pegasi	Markab	2.46				45
Ṣāra (*al-akhīra*)*	α Andromedae	Alpheratz	2.06	20	20		20
divider	γ Pegasi + δ Andromedae	Algenib (γ Peg)	2.81 3.25				20
Sa^{c}d	β Andromedae	Mirach	2.06	35	35		45
divider	γ Andromedae	Almach	2.09				20
	β Trianguli		3.00				5-7
Chūchabayn	α + β Arietis	Hamal + Sheratan	2.62 2.00	25	25	40	30
	α + β Persei	Mirfak + Algol	1.78 2.06				27
	δ Persei		3.00				15
	ε Persei		2.87				15
	ζ Persei	Atik	2.81				10
Thurayyā	Messier 45	the Pleiades	>=2.84	30	30		30
divider	λ Tauri		3.40				30
Debrān	α Tauri	Aldebaran	0.84	35	35		32.5
divider	β Tauri	Elnath	1.62				15
divider of the divider	θ Aurigae	Menkalin	2.62				22.5
Shubcha	λ + φ^1 + φ^2 Orionis	Meissa (λ)	3.37 – 4.37	20	30		20
divider	α Orionis	Betelgeuse	0.43				30
Jawza	γ Geminorum	Alhena	1.90	40	35		40
divider	β Geminorum	Pollux	1.15				35
Sha^{c}ara	α Canis Minoris	Procyon	0.37	30	30		30
divider	β Cancri	Altarf	3.50				30
Janb	δ, ε + ρ Hydrae		3.37 – 4.34	40	30		40
divider	β Ursae Majoris	Merak	2.31				30
Dharā^{c}īn	α + γ Leonis	Regulus + Algieba	1.34 2.00	30	25		30
divider	γ Ursae Majoris	Phecda	2.40				25
Buṭayn	δ Leonis	Zosma	2.50	20	20		40
Sufā^{c}	β Leonis	Denebola	2.12	30	50		30
divider	α^2 Canum Venaticorum	Cor Caroli	2.84				50
Mīyathāb	ε Virginis	Vindemiatrix	2.84	30	30		60
Dhakarayn	α Boötis	Arcturus	-0.07	20	20	30	20
divider	ε Boötis	Izar	2.34				20
divider	δ Boötis		3.43				30
Ghafur	α Coronae Borealis	Alphecca, Gemma	2.21	45	45		40
divider	β + γ Herculis	Kornephoros (β)	2.75 3.71				20
divider	β Draconis	Rastaban	2.78				30
^{c}Aqrab	γ Draconis	Eltanin	2.21	30	20	20	40
divider	α Scorpii	Antares	1.03				5
divider	α Ophiuchus	Rasalhague	2.06				25
* *awliyya*, and elsewhere *ūlā* means "first"; *akhīra*, and elsewhere *akhīr*, means "last".							

Table 6.7 ***Falaj*** **stars identified in Qarya and their timings**

On the whole, the divider stars are selected to mark divisions between *athar*s, but in some cases their timing seems to bear little relation to the *athar* divisions. For example, between Kuwī and Mūfī there are three *athar*s, the first two each 40 minutes and the third 35 minutes long, but timing between the stars is different, as shown in Figure 6.11, with four official and one (γ Cygni) unofficial dividers.

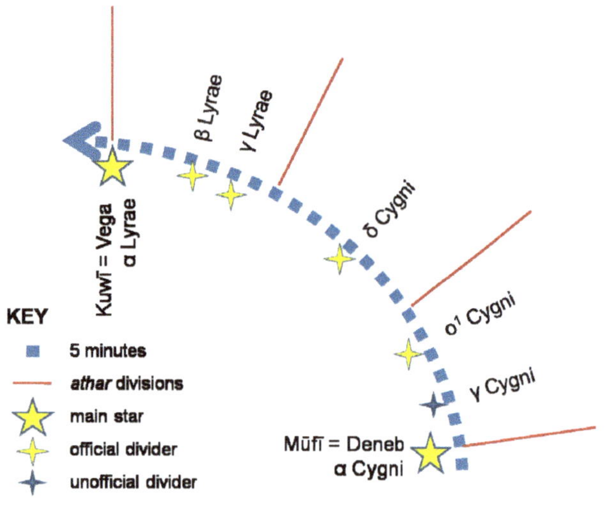

Figure 6.11 Example of star and *athar* divisions in Qarya: Kuwī to Mūfī

Several stars are used in addition to those listed in Table 6.7, either as subdividers or as markers when other stars are not clear. They include Capella (α Aurigae), here called *ᶜayyaq* ath-Thurayya, which rises eight minutes before Aldebaran, and the three stars of the belt of Orion called Mīyazīn, which rise 10, 15 and 20 minutes after Shubka, with the third (Alnitak, ζ Orionis) rising at the same time as the official divider, Betelgeuse.

The locations of the stars listed in Table 6.7 and mentioned above are shown in Figure 6.12. These views of the longest and shortest nights of the year cannot be seen by man, but present the whole sky from horizon to horizon so that all of the stars used throughout the year can be shown. On the shortest night, i.e. mid-summer, Mūfī (Deneb, α Cygni) is currently the first main star to rise and Debrān (or Dabarān - Aldebaran, α Tauri) the last; on the longest night, i.e. mid-winter, Shubka (λ, φ^1 + φ^2 Orionis) is the first and Kuwī (Vega, α Lyrae) the last.

6.4.2 Stars formerly used in Misfa

The twenty-one main stars and some of the dividers formerly used in Misfa were identified (Table 6.8), some from looking at the night sky, but the majority from Starry Night™.

The similarity in star names in Misfa and Qarya is to be expected, and those with the same names are generally the same star. However, in Misfa, Debrān refers to Capella not to Aldebaran, which although also called Debrān is not used for the *falaj*.[43] Nubtā is a variation of Baṭn and is also Zosma (δ Leonis). Nijam could be a variation of the name Janb, and unlike Qarya seems to include all of the stars in the head of Hydra. Hamal is not part of Chowchabayn because it rises seven minutes later than Sheratan, and is replaced by a closer star, Mesarthim (γ Arietis); given the meaning "the two stars" for this marker, it appears that it is not an option to use only one.

Star name in Misfa*	Classification	English name
Chowchabayn	β + γ Arietis	Sheratan + Mesarthim
Thurayya	Messier 45	Pleiades
Dabrān/Debrān/ Dibrān	α Aurigae	Capella
divider	β Tauri	Elnath
Shabcha	λ, φ^1 + φ^2 Orionis	Meissa (λ)
Mīyazīn (divider)	δ, ε + ζ Orionis	Mintaka, Alnilam + Alnitak
Juzā	γ Geminorum	Alhena
divider	α Geminorum	Castor
Shaᶜara	α Canis Minoris	Procyon
divider (Suhayl)	α Carinae	Canopus
Nijam/Janb	stars in Hydra	head of Hydra
Dhirāᶜayn	α + γ Leonis	Regulus + Algeiba
Nubtā/Baṭn	δ Leonis	Zosma
Sufāᶜ	β Leonis	Denebola
Mawāfī/Mīyathāb	ε, δ + γ Virginis	Vindemiatrix, Auva + Porrima
Dhakarayn	α Boötis	Arcturus
divider	ε Boötis	Izar
Ghafar	α Coronae Borealis	Alphecca, Gemma
ᶜAqrab	β, δ + π Scorpii	Graffias, Dschubba + 1
divider	α Scorpii	Antares
Chuwī	α Lyrae	Vega
Mūfī	α Cygni	Deneb
Ghurāb	α, β, δ + λ^2 Delphini	Sualocin, Rotanev + 2
Adam	ε Pegasi	Enif
Ṣāra (*al-ūlā***)	β Pegasi	Scheat
Ṣāra Marra Thāniyya**	α Andromedae	Alpheratz
Saᶜd	β Andromedae	Mirach
* mainly as spoken by Masud and Khalfan Al-Hattalli; ** *ūlā* means "first", *thāniyya* means "second", *marra* means "time".		

Table 6.8 Stars formerly used for time keeping in Misfa

6.4.3 Stars formerly used in Al Hamra

A modern schedule of main stars for Al Hamra is given in Appendix A.2, and the relevant information in Table 6.9. Unlike lists from other villages, a description of the stars is given. Badr al-ᶜAbrī also provides a "generic" star list (c. 1980: 30), considered to apply to Al Hamra where he lived. The names he gives are the same as in Table 6.9 with minor differences in spelling.

[43] In Barzaman, Aldebaran and Capella are also treated as one and called Bū-Qābil (see Chapter 7).

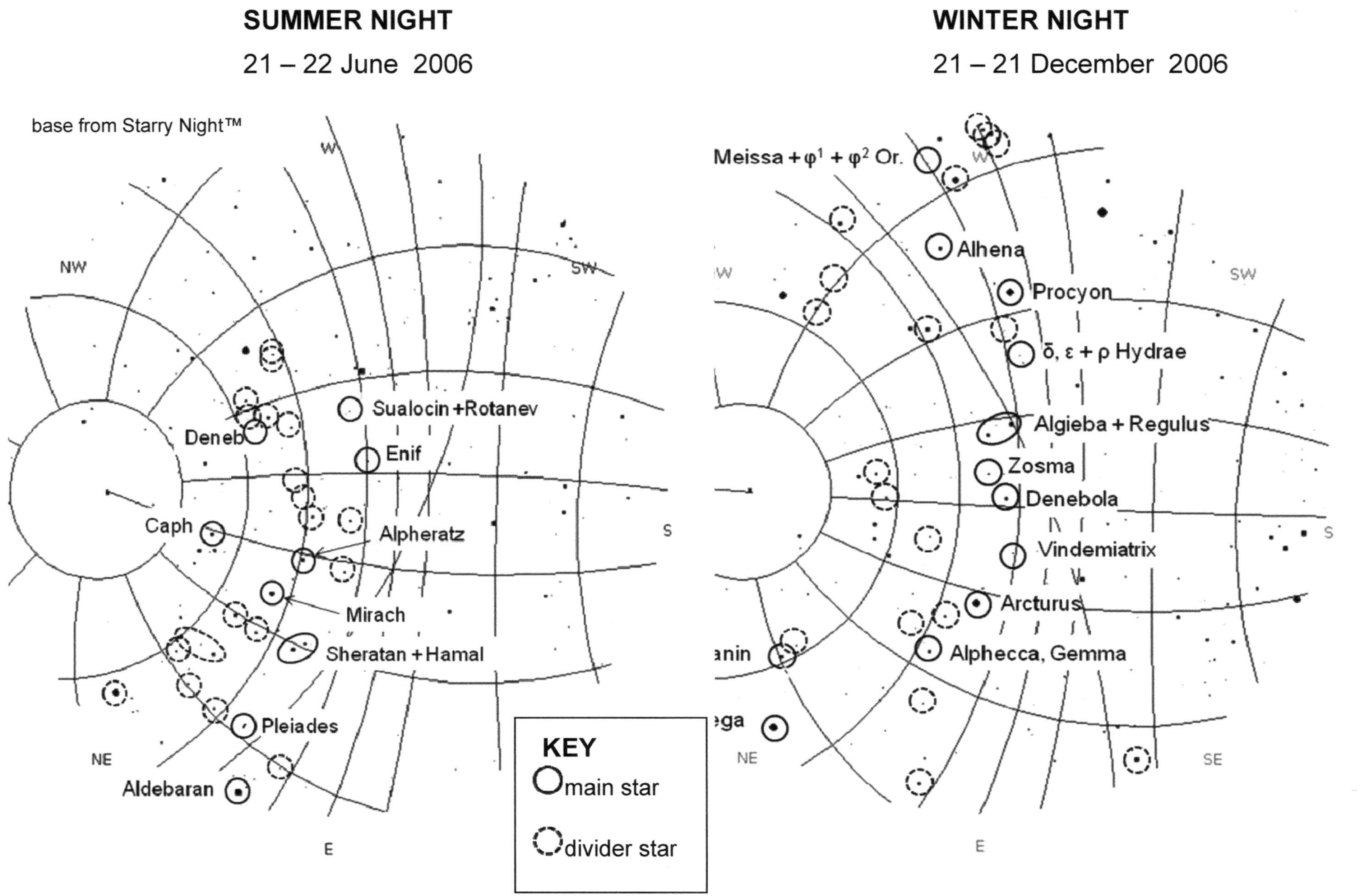

Figure 6.12 *Falaj* stars indentified in Qarya

Star name from modern schedule		Description	Start (d/m)	Number of		
Arabic	Transliteration			*athars*	days	min
الثريا ء	Thurayyā'	close cluster	12/5	2	16	60
الدبران	Dabarān	1 bright	28/5	2	16	65
الشبكة	Shubka	close cluster, small stars	13/6	2	16	50
الجوزاء	Jawzā'	1 bright	29/6	2	16	75
الشعرة	Sha^c^ara	1 bright	15/7	2	16	80
الجنب	Janb	close cluster	31/7	2	16	55
الذراعين	Dhirā^c^ayn	2 joined rising at the same time	16/8	2	16	55
البطين	Buṭayn	1 single bright	1/9	2	16	40
السفاع	Sufā^c^	1 single bright	17/9	2	16	80
الميثاب	Mīyathāb	3 spread out	3/10	2	16	60
الذكرين	Dhakarayn	2 bright	19/10	3	24	68
الغفر	Ghafr	1 bright	12/11	2	16	90
العقرب	^c^Aqrab	group, spread out	26/11	3	24	70
كوي	Kuwī	1 bright	20/12	3	24	115
الموفي	Mūfī*	2 bright	13/1	2	16	40
الغراب	Ghurāb	cluster, small stars	27/1	2	16	15
الأدم	Adam	1 small	10/2	2	16	45
الصارة الأولى	Ṣāra l-Ūlā	1 medium bright	24/2	2	16	65
الصارة الاخيرة	Ṣāra l-Akhīra	1, brighter than the first	11/3	2	16	40
السعد	Sa^c^d	1 bright	27/3	2	16	70
الكوكبين	Kawkabayn	2 bright, close together	12/4	3	24	100

*also referred to as *al-ṭayr* (bird) by Abdullah Al-Abri

Table 6.9 Modern schedule of main stars formerly used in Al Hamra

The details in the Al Hamra schedule are similar to those from Qarya, but with different start dates and minor differences in the spelling of star names. The main factors affecting the date of rise of a particular star are the time at which the information was relevant, the longitude, and the shape of the horizon. Since the stars rise four minutes earlier each day, a difference in the start date of a full day is equivalent to approximately one year in the date of compilation. Al Hamra is a couple of kilometres to the west (Figure 5.1), and the stars rise a little later than in Qarya. The data provided could, in theory, give information on when the lists were compiled. However with differences in start dates for the stars ranging from one to ten days, a simple correlation cannot be made.

In addition to the star list in Table 6.9, descriptions of the *falaj* stars for Al Hamra are given by Badr al-^c^Abrī (c. 1980: 32-41) from an "old" manuscript by Sheikh Sa^c^īd b. Sāliḥ al-^c^Abrī. The descriptions cover the main stars and their dividers, including their relative brightness, location of rise relative to other stars and other markers used for clarification, such as stars setting or in the zenith. With this information it was possible, with the aid of Starry Night™, to identify a number of the stars with a fair degree of certainty, taking care not simply to assign the same identification as in Qarya or Misfa.

Only 19 stars are described in the book, whereas 21 would be expected in this area. On a night when Mawāthīb (Mīyathāb in the Table) is used, in the book there would be a gap of approximately 3½ hours to ^c^Aqrab, the next main star, too long for the two *athar*s represented by Mawāthīb. One of the missing stars was almost certainly Dhakarayn, which is referred to in the description of Mūfī, another main star. Badr al-^c^Abrī was too ill to discuss his work, but I was taken by Harith bin Nabhan Al-Abri (*wakil* of the *falaj* at Qala) to meet Abdullah bin Mohammed Al-Abri (29 October 2007). As well as giving me the modern star list, he said that the author of the manuscript, Sheikh Sa^c^īd b. Sāliḥ, had died in 1921, and that his (Abdullah's) father, Muḥammad b. ^c^Āmar al-^c^Abrī, had written a manuscript on stars some 60 years ago, i.e. c. 1947. This turned out to be a copy of Sheikh Sa^c^īd's work: it has similar, though not identical, wording to that in Badr al-^c^Abrī's book, and being complete, it also has information on the two main stars missing from the book: Dhakarayn and Ghafr. The first page is shown in Figure 6.13., and has a section on Dhakarayn, described as two bright stars some distance apart, the northern (*na^c^shī*) of which is used to mark (the beginning of) three *athar*s (to the next star). Throughout the manuscript, the terms *na^c^shī* and *suhaylī*, meaning "of the Plough" and "of Canopus", are used to describe north and south respectively. This is helpful, as the direction is clear: the use of *shimāl* for north can be misleading as it can also mean "to the right", which is south when facing east.

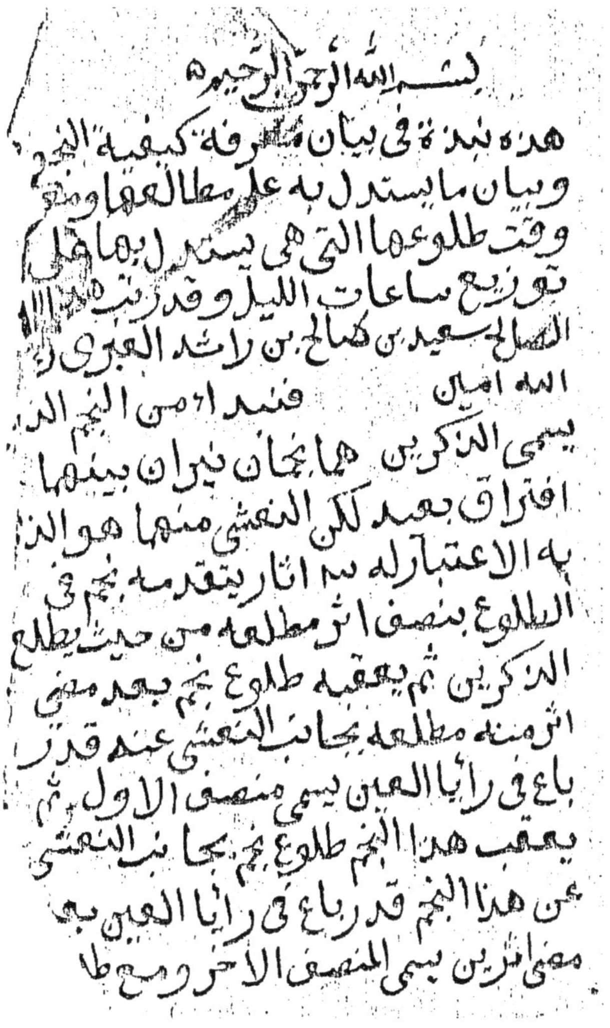

Figure 6.13 Folio 1 from manuscript on stars used in Al Hamra
copyist Muḥammed bin ᶜĀmar al-ᶜAbrī c. 1947

This is not the same version of the manuscript reported by Badr al-ᶜAbrī who starts with ᶜAqrab, which in this version is on the second page, following a complete description of Ghafr. If Badr al-ᶜAbrī had been using this version, he would have seen the information for Ghafr and included it. Also, the manuscript held by Abdullah bin Mohammed Al-Abri does not give hours and minutes for times, only *athar*s; distances are not given in metres but mostly in *bāᶜ*, and the spellings of some star names differ. Either Badr al-ᶜAbrī added the more modern measurement systems, or he was using a more recent version of the manuscript. A check on the distances given between stars (Appendix A.3) shows that a *bāᶜ*, given as 1.5 m by al-ᶜAbrī, is equivalent to 6° to 8° of declination (latitude).

With the complete manuscript, albeit only a photocopy, it was possible to firmly identify more stars, including those for which the artificial timing of ½ hour for one *athar* given by Badr al-ᶜAbrī did not fit with the description, and those for which the description in the book was, apparently, in error. One example is ᶜAyyaq al-Dabarān, which in the book (pg 37) is said to be brighter than Dabarān (Aldebarab has a magnitude 0.84), and to rise ¼ *athar* before and to the south of Kuwī (Vega, α Lyrae), i.e. south of 38°47'N. In the manuscript, this is ᶜAyyaq, again brighter than Dabarān but rising north of Kuwī. From the latter, there is no doubt that the star referred to is Capella (α Aurigae), which has a magnitude of 0.06 (i.e. brighter than 0.84) and is at 40°0'N. The stars are listed in Table 6.10, with the Arabic taken from Muḥammad b. ᶜĀmar al-ᶜAbrī's copy of Saᶜīd b. Sāliḥ al-ᶜAbrī's manuscript. Particular care was taken over star names used for more than one star: for example, Dabarān here is definitely Aldebaran (α Tauri), being described as rising south of the Pleiades, whereas Capella (α Aurigae) rises north of the Pleiades. Provisional identifications are given in italics and the details of the star descriptions including comments on their identification in Appendix A.3, together with calculations of distances between the rising points of some stars, used to help with their identification.

The descriptions in the manuscript are consistent with, though more detailed than, those given in the modern list in Table 6.9. There are several other stars in the manuscript that could not be identified, which are mostly indicators or "unofficial" dividers.

6.4.4 Stars formerly used in Qala

A star list for Qala, said to have been written about 100 years ago, and similar to the modern schedules for Qarya and Al Hamra was supplied by Harith bin Nabhan Al-Abri in November 2006. The full list is given in Appendix A.4, and the relevant information is given in Table 6.11

Star name		Classification*	English name*	Star name		Classification*	English name*
Arabic	Transliteration			Arabic	Transliteration		
		η Boötis	Muphrid			*ε Persii*	
الذكرين	Dhakarayn	α Boötis + α Virginis	Arcturus + Spica	الثرياء	Thurayyā'	Messier 45	Pleiades
	1st divider	ε Boötis	Izar			*37 Tauri*	
	last divider	δ Boötis			divider	λ Tauri	
الغفر	Ghafr	α Corona Borealis	Alphecca	العيق	ᶜAyyaq	α Aurigae	Capella
		η + ζ Herculis		الدبران	Dabarān	α Tauri	Aldebaran
العقرب	ᶜAqrab	β + δ Scorpii + *other(s)*	Graffias, Dschubba		divider	β Tauri	El Nath
	divider	α Scorpii	Antares			β Aurigae	Menkalinan
كوى	Kuwī	α Lyrae	Vega			γ Orionis	Bellatrix
	divider	δ Cygni		الشبكة	Shubka	λ, $φ^1$ + $φ^2$ Orionis	Meissa (λ)
		γ Cygni	Sadr		divider	α Orionis	Betelgeuse
الموفى	Mūfī	α Aquilae + α Cygni	Altair + Deneb	الجوزة	Jawza	γ or ε Geminorum	Alhena or Mebsuta
		ε Cygni	*Gienah*		divider	α Geminorum	Castor
الغراب	Ghurāb	α, β, δ + λ2 Delphini + *others*	Sualocin (α), Rotanev (β)			β Geminorum	Pollux
	divider	1 Pegasi		الشعرة	Shaᶜra	α Canis Minoris	Procyon
الادم	Adam	ε Pegasi	Enif		divider	*β Cancri*	*Altarf*
	divider	θ Pegasi	Baham	الجنب	Janb	δ, ε + ρ Hydrae + *others*	head of Hydra
		η Pegasi	*Matar*			α Ursae Majoris	Dubhe
الصارة لأولى	Ṣāra l-Ūlā	β Pegasi	Scheat		divider	*μ or ε Leonis*	
		β Cassiopeiae	Caph			β Ursae Majoris	Merak
		α Pegasi	Markab			ζ + η Leonis	Aldhafera (ζ)
		α + γ Cassiopeiae	Schedar (α)	الذراعان	Dhirāᶜān	α + γ Leonis	Regulus, Algieba
الصارة الثانية	Ṣāra l-Thāniyya	α Andromedae	Alpheratz		divider	*ρ Leonis*	
	divider	*δ Andromedae*	*Ruchbah*			δ Ursae Majoris	Megrez
السعد	Saᶜd	β Andromedae	Mirach	البطن	Baṭn	δ Leonis	Zosma
	divider	γ Andromedae	Almach			ε Ursae Majoris	Alioth
		α Trianguli	Rasalmothallah	السفاع	Sufāᶜ	β Leonis	Denebola
الكوكبين	Kawkabayn	α/γ + β Arietis	Hamal/Mesarthim + Sheratan			ζ Ursae Majoris	Mizar
		β (+ *κ + α*) Persii	Algol (β)		divider	$α^2$ Canum Venaticorum	Cor Caroli
		41 Arietis				η Ursae Majoris	Alkaid
				المواثيب	Mawāthīb	ε, δ + γ Virginis	Vindemiatrix, Auva + Porrima

* italics: identification uncertain.

Table 6.10 Identification of stars formerly used in Al Hamra

Star name		Start (d/m)	Duration (d)	Comment
Arabic	Translit.			
الموفي	Mūfī	15/1	15	very cold
الغراب	Ghurāb	30/1	16	
الادم	Adam	16/2	16	
الأولى الصارة	Ṣāra l-Ūlā	2/3	16	
الاخير الصارة	Ṣāra l-Akhīr	17/3	16	
السعد	Sa^{c}d	3/4	16	
الكوكبين	Kawkabayn	18/4	24	
الثريا	Thurayyā	12/5	16	
الدبران	Dabarān	28/5	15	
الشبكة	Shubka	12/6	16	return of the sun; change to high summer 22/6
الجوزه	Jawza	26/6	16	
الشعرة	Sha^{c}ara	12/7	15	
الجنب	Janb	26/7	16	
الذراعين	Dhirā^{c}ayn	12/8	16	
لبطن	Baṭn	28/8	16	
السفاع	Sufā^{c}	13/9	16	
الميثاب	Mīyathāb	28/9	16	
		14/10	10	no star
الذكرين	Dhakarayn	24/10	23	
الغفر	Ghafr	16/11	15	return of the sun; transition to deep winter
العقرب	^{c}Aqrab	1/12	23	
كوي	Kuwī	23/12	23	

Table 6.11 Main stars formerly used in Qala

Falaj star	Name	Paired star	Name
Qarya			
Mūfī	Deneb	Mūfī (*al-janūbī*)*	Altair
Ṣāra *l-awliyya*	Caph	Ṣāra *l-awliyya* (*al-janūbiyya*)*	Scheat
Ṣāra *l-akhīra* (*al-janūbiyya*)	Alpheratz	Ṣāra *l-akhīra*	Ruchbah
Sha^{c}ara (*al-shimālī*)*	Procyon	Sha^{c}ara (*al-janūbī*)*	Sirius
Dhakarayn (*al-^{c}ālī*)*	Arcturus	Dhakarayn	Spica
^{c}Aqrab	Eltanin	^{c}Aqrab (*al-janūbī*)*	Graffias + Dschubba
Misfa			
Mūfī	Deneb	Mūfī	Altair
Sha^{c}ara	Procyon	Sha^{c}ara	Sirius
Debrān	Capella	Debrān	Aldebaran
Al Hamra			
Dhakarayn	Arcturus		Spica
Mūfī	Deneb		Altair
Sha^{c}ara (*al-na^{c}shiyya*)*	Procyon	Sha^{c}ara (*al-suhayliyya*)*	Sirius
* *janūbī, janūbiyya* = south; *shimālī* = north; *^{c}ālī* = high (north in areas on the south side of the mountains); *na^{c}shiyya* = north, *suhayliyya* = south. Star identifications are given in previous tables.			

Table 6.12 Star pairs in the Qarya area

6.5 STAR PAIRS, CONSTELLATIONS AND STAR LORE IN THE QARYA AREA

Several of the star names used in the Qarya area refer to more than one star, often a pair, which may be a long way apart. The two stars in both Kawkabayn and Dhirā^{c}ayn are relatively close and are generally considered as one, i.e. the names are not qualified as being high/low, north/south. Other pairs identified during the study are given in Table 6.12.

In Qarya, the northern star is used officially to mark the *athar*s, with the exception of the second Ṣāra. For Al Hamra, the northern star was also used. However, the list is incomplete: in Misfa the southern stars of star pairs were used preferentially because the mountains restrict the view to the north, and the three on the list are the only ones for which the northern star was used. This means that there were more pairs, probably similar to those in Qarya, but with the southern star used.

Only two of the stars in Scorpius are known as ^{c}Aqrab (the Scorpion) in Qarya: the constellation known here does not include Antares (α Scorpii), which is, however, used as a divider star. Mīyathāb (Mawāthīb) deserves special mention: it is seen in Qarya as a constellation of seven stars, but only one, Vindemiatrix (ε Virginis), is used officially for time keeping. The stars are shown in Figure 6.14: the first, Muphrid (η Boötis), is close to Arcturus (α Boötis), and all seven together can be seen to reflect the meaning of the name as something leaping, somewhat running out of steam by Minkar (ε Corvi). In the south of this constellation, the stars Algorab, Gienah, Minkar and Kraz (δ, γ, ε and β Corvi) are the brightest in the international constellation Corvus, the Crow, in Arabic Ghurāb. This bears no relation, except perhaps for the quadrilateral shape of the four main stars, to the *falaj* stars called Ghurāb in Qarya and other study villages, which are in the constellation Delphinus.

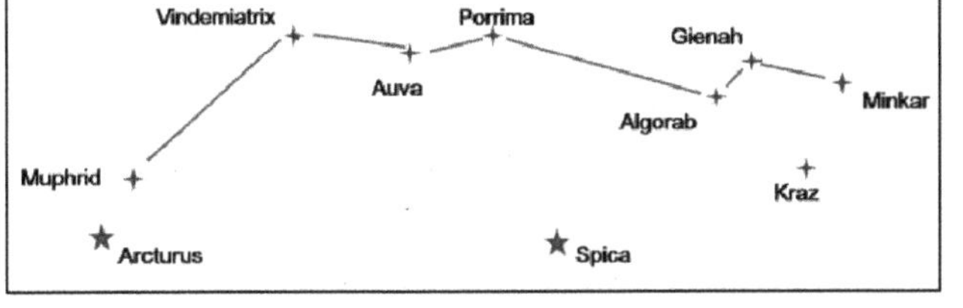

Figure 6.14 Mīyathāb (leaping) in Qarya
(facing east)

While recording star names, some information on agricultural activities associated with certain stars in Qarya was also collected:

Recording (11 January 2006)	**Rough translation**
an-nakhala alḥīn – chuwī	date palms – now – Chuwī, i.e. Chuwī for (planting) date palms
al-ᶜaṣīr fī l-mūfī - ᶜaṣīr sajjar	juice is in Mūfī – juice from sugar cane
l- ghurāb – ṣayf	Ghurāb - summer
tārīkh thūm fī l-mīyathāb	the time for garlic is in Mīyathāb
tārīkh ṣayf adh-dhakarīn; wa l-mwāsim kadhālik – mawsim dhakarayn – maṭṭar	the time of summer is in Dhakarayn; and the season of Dhakarayn - rain
tārīkh adh-dharraᶜ – waqt/jaww ḥār – fī d-debrān. hādha tārīkh aṣ-ṣayf	the time for wheat – hot weather – in Debrān, this is summer time
tārīkh al-jizlān - shaᶜrā'	the time of plenty is in Shaᶜrā'

When Chuwī, for example, is the last star at night, i.e. to be seen rising at dawn, that is "*waqt* or *tārīkh* Chuwī", the time of Kuwī, as for the *falaj* schedule. The planting of date palms was not in the recording, but was demonstrated at the time. Other sayings were recorded on the same evening, but with several people talking at the same time, they are not easy to decipher, and the associations for Kuwī and Mūfī could not be heard. Each begins with *wa ṭalaᶜ fī* or *yaṭlaᶜ fī* (referring to sunrise, and specifically its heliacal rising) followed by the star name and then a rhyming association:

yaṭlaᶜ fī l-ghurāb jurāb	when it rises in Ghurāb – bags full of dates. (*Jurāb* is a bag woven from date palm leaves and full of dates)
wa yaṭlaᶜ fī s-saᶜd -juzz al-jaᶜd	if it rises in Saᶜd, it is the time of shearing
lam yaṭlaᶜ fī d-dabrān ākhar layl al-qayr[44] *fī kull al-awṭān*	when it rises in Dabrān, there are dates in all (Arab) states

Following this, a few somewhat different sayings were written down by one of Rashid Khamis's family (9 May 2006):

Arabic	**Rough translation**
بطالِع الموفي ثَجي جراب وتوفي	the rise of Mūfī brings bags full of dates and plenty (prosperity)
بطالِع الغُراب ثَجي جراب	the rise of Ghurāb brings bags full of dates
بطالِع الأدم ماتَنْدَم	the rise of Adam – you won't regret
بطالِع الصارة عَصارة	the rise of Ṣāra is the time for (sugar cane) juice
بطالِع السَعْد قوْقد[قرفد] الجَعْد	the rise of Saᶜd is the time for shearing

The above are all *falaj* stars, and the list is probably not complete. In Qala, according to Harith Al-Abri (28 January 2006), Suhayl (usually Canopus, α Carinae) was used in the same way, as an indicator for the cultivation of some plants.

The similarity in star names in Qarya, Qala, Misfa and Al Hamra is to be expected, being so close together. Misfa and Al Hamra probably have the closest link, since Saeed Ali Al-Khiyari, a farm worker in Al Hamra says (3 April 2007) that people from one own water shares in the other. It is possible that one settlement learned from another, with Qarya probably being the first to use this system of timing with stars.

The variation in the endings of some star names has a significant effect on the meaning. However, most people do not think about the meanings of star names, a notable exception being Sultan Al-Shidhani in Al Fath who describes Shābik as its dictionary definition: the fishing net. It appears that there is no right or wrong, and the endings are given here as provided by a particular source. Elsewhere, the spellings given in published works by astronomers or academics (e.g. Badr, 1988) are used.

44 Corrected from *khayr* (well-being) following clarification given in October 2009.

7 BARZAMAN

7.1 INTRODUCTION

In Barzaman there is one main tribal group, the Aamari. According to al-ᶜAbrī (c. 1980: 77-78), the *falaj* was built in Yaᶜāriba times, in the time of Imam Sulṭān b. Sayf (d. 1123/1711), but was not completed to yield water for irrigation until the time of Turkī b. Saᶜīd, who was Imam from 1287/1871 to 1305/1888. However, the *wakīl* (administrator) of the *falaj*, Salam bin Abdullah bin Dhiyab Al-Aamari, is certain that it was operational before that time. Moreover, Sheikh Sultan bin Mohammed Al-Habsi in Al Fath, who is well informed about the history of the area, says that Barzaman is older than Al Fath, i.e. over 350 years old,[46] and given that a *falaj* is the most essential requirement for agricultural settlement in this area, it appears that Barzaman and its *falaj* were indeed built in the 11th/17th C.

The layout of the village is shown on Figure 7.1. An area of fields formerly cultivated by the *falaj* can be seen to the south of the current area of cultivation. Apparently this was abandoned in c.1973-1975 when the *falaj* flow became insufficient for the seasonal cultivation carried out there.[47] It was in the same period that the new village was developed.

The flow of water is said to be sufficient for the area irrigated now, even though it reduces when a well (or wells) some 15 km from the *falaj* source is pumped by the Municipality for drinking water. The reduction is said to be noticeable a few hours after pumping the well, consistent with hydrogeological principles. It appears from the hydrograph (Figure 5.4) that the flow has gradually decreased since regular monitoring began in 1990; this can be attributed to a decline of water levels in the source area due to abstraction from pumped wells.

The first use of the *falaj* water today is in the *mughsila*, the building for washing corpses (Figure 7.2), close to but downstream of the *sharīᶜa* (the point of first main access for domestic use). Currently, the water is pumped from the *falaj* to a tank on the roof, but previously it flowed through the building under gravity. This is followed by toilets, offset from the *falaj* channel, and then wash rooms usually attached to mosques. Many of these are still used, as is the main *falaj* channel, at least for washing clothes. When everyone lived in the old village, some 500 m south of the new town, although there was irrigation upstream, the water was clean and drinking water was taken from near the houses, not from the *sharīᶜa*.[47]

7.2 ORGANISATION OF THE *FALAJ* AT BARZAMAN

The main *falaj* officials are listed in Table 7.1. There are two *wakīl*s (administrators), so that people have someone to go to if one is away. Both are also the two *ᶜarīf*s (officers) for the *falaj*. Two others, Suleiman bin Saeed and Humaid bin Ali are responsible for the day-to-day distribution of water, and Suleiman is also the auctioneer for the *falaj* water. They organise the distribution of water to be as convenient as possible. For example, in one irrigation cycle Hamud bin Saeed bin Salim has one *athar* at 7 pm and another at midnight, but Suleiman arranges the distribution so that Hamud and others can take their water in one block.[48] Decisions regarding the *falaj*, such as the settlement of disputes and selection of officials, are made by a group of village elders of about ten to twenty people.

The irrigation cycle (*dawrān*) is nine days, and the day starts at sunset and the *bādda* (a variation of *badda)* is 48 *athar*s, i.e. one full day. The *badda* names are:

al-jumᶜa	*as-sabt*	*al-qāᶜida*
Friday	Saturday	auction day
al-ithnayn	*la-muzayyida*	*ath-thalātha*
Monday	additional day	Tuesday
al-quᶜid	*al-arbaᶜa*	*al-khamīs*
auction day	Wednesday	Thursday

Although most are named by days of the week, they will only fall on that day from time to time. Terms with the root √ (*q.ᶜ.d*) when related to *aflāj*, generally mean the auction of water for *falaj* maintenance, and could have their origin either in the common meaning of sitting, in that people do sit around at an auction, or in the meaning of *qāᶜida*, a foundation, in that this was the basis for sustainable management of the *falaj*. According to Wilkinson (1977: 112-114), the *qāᶜida* was inalienable and could not be changed whereas the *muzayyida*, also for running the *falaj*, was more flexible and did not have to be used. Today, it appears that these terms are simply names, except for that actually used to raise money for the *falaj*, in this case the *quᶜid*.

Each 24-hour day has four divisions, and people's share changes from one irrigation cycle to the next in the order shown in Figure 7.3.

---

[46] 20 October 2007.

[47] Salam bin Abdullah Al-Aamari, 16 October 2007.

[48] Hamud bin Saeed, 1 May 2006.

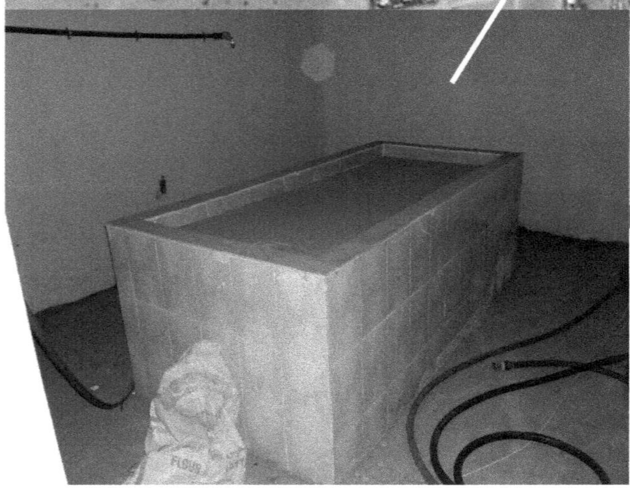

Figure 7.2 The *mughsila* at Barzaman October 2007

Figure 7.1 Layout of Barzaman

Title	**Name** (all Al-Aamari)	Role	Started	Other work
wakīl and *ᶜarīf*	Salam bin Abdullah bin Dhiyab	administration, accounts, supervision of works and repairs	c. 1997	Ministry of Awqāf in Ibra
	Saif bin Hamad			teacher of the Holy Qur'ān in Barzaman
dillāl	Suleiman bin Saeed bin Saif	auctioneer, distribution of water	1950s-1960s	farmer
	Humaid bin Ali	distribution of water	1950s-1960s	farmer

Table 7.1 ***Falaj* officials in Barzaman**

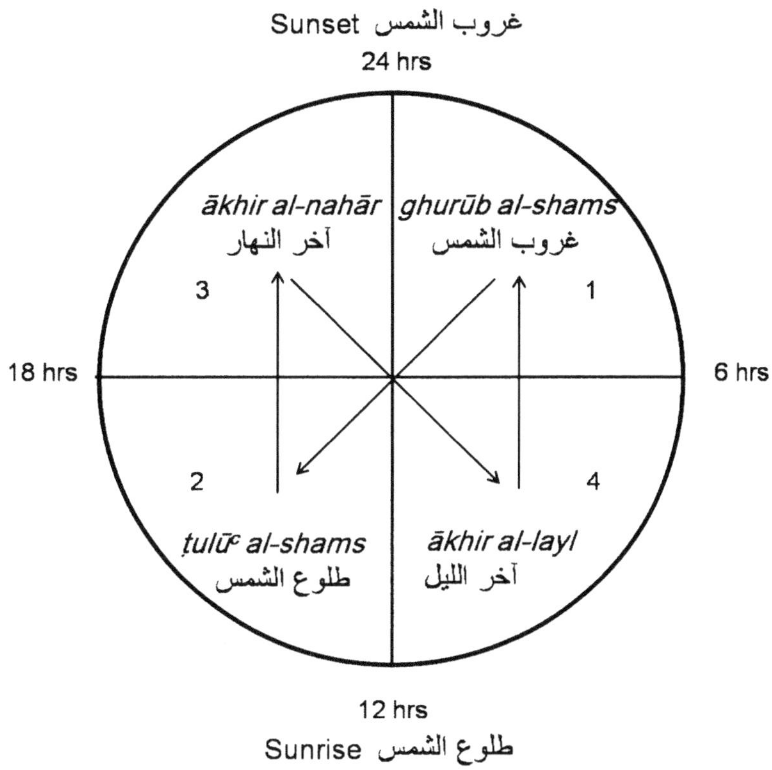

Figure 7.3 Rotation of water shares in Barzaman

A page of the *wakīl*'s notes (Figure 7.4) gives the division of water following sunset for *raddat al-sabt*, indicating that a *radda* here is the same as a *bādda*. The time is divided into *qaḍiyān* as follows:

Sunset: for Mohammad bin ᶜAlī Wadūnī, 2½ *athar*s

from *qaḍiyān*, 2 *athar*s less a part. to *qaḍiyān* 3 less a part for Farīsa bint ᶜAlī, 1 (?+ ½) *athar*s;

from *qaḍiyān* 3 less a part to *qaḍiyān* 5 ...for Masᶜūd ᶜAbdallāh, 1½ *athar*s;

from *qaḍiyān* 5 to *qaḍiyān* 3 before *b.t.ᶜ* for Ḥumayd bin ᶜAshᶜīsh ...;

from *qaḍiyān* 3 before *b.t.ᶜ* to *qaḍiyān* 10 for Ḥumayd bin ᶜAlī (2 *athar*s +?).

Qaḍiyān is a term used in Barzaman to name the *athar*s in the first part of the night, at least in written documents. Following the *athar*s associated with sunset, there are ten *qaḍiyān athar*s up to the middle of the night. After this, the *athar*s are numbered *min al-layl* (from the night); the first part of the day is called *rukhṣān*, meaning "soft, tender, low-priced or authorised" and the second *ruwāḥ*, meaning "departure".[49] The clearest explanation given of this system of naming *athar*s was from Mudaybi, and is given in Appendix B; however it is unclear why the four quarters of the day are given two sets of names: that described here and that given in Figure 7.3.

[49] Salam bin Abdullah bin Dhiyab, April 2008.

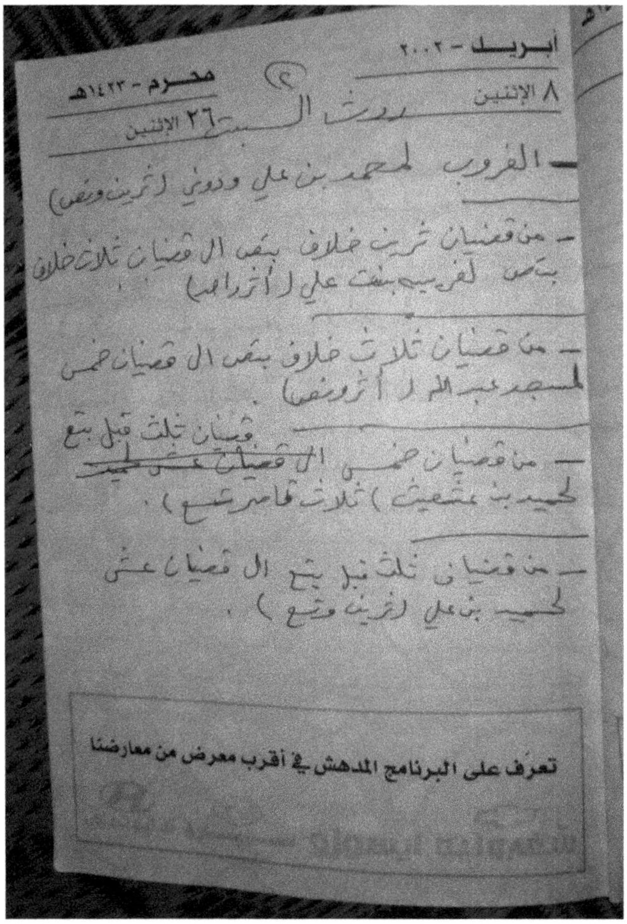

Figure 7.4 The *wakīl*'s notes, Barzaman

Time keeping with the stars and sun for the allocation of *falaj* water is said to be a very old custom, and the water clock (*ṭāsa*) is not known by the *wakīl*. Watches were available before the 1970s from the Emirates, Saudi Arabia and Zanzibar, where many people have connections. Apparently most, but not all, of the expatriate farmer workers use watches at night, and from my visits it appears that this is also the case for many Barzamanis.

Figure 7.5 Sundial at Barzaman (30 March 2007)

The sundial (Figure 7.5) is in the old village in the middle of the area of irrigation; the grooves in the ground are *athar* divisions with times ranging from 20 to 40 minutes. Some people have marked their time with date palm leaves and small stones. The dial is said to be used by most farmers,[50] and this seems to be the case given the number of personal markers.

Little information was collected on water ownership in Barzaman. The major shareholders are:

Salim bin Al-Abd 20 *athar*s
Hamdun bin Abdullah 19 *athar*s
Salim bin Humaid 10 *athar*s.

Most people own between five and seven *athar*s, while Salam and his family own three.

Saeed bin Abdullah bin Dhiyab, Salam's brother, has at least six documents on water rights given to him by his uncle and thought to be over 100 years old.[51] A photograph of one is shown in Figure 7.6. They all give much the same information: who sold how much water to whom, for how much, and which *radda* or *badda* (here *bādda*) the water is from. The *radda*s specified in these documents are *al-muzayyida, al-sabt, al-arbaᶜā* and *al-q.ᶜāda* confirming that here, a *radda* is the same as a *badda*, and that the names have not changed since these documents were written.

A summary of the *falaj* finances is given in Table 7.2. One day of the cycle (*al-quᶜid*) is for *falaj* maintenance and auctioned every cycle. This raises approximately 48 riyals each time, possibly more in summer, giving a total of approximately 1840 riyals each year. Additional sources of funds are the four *athar*s owned by the *falaj*. Every year additional funds are raised in the month of Rajab,[52] when everyone pays 1 riyal per *athar* as a sort of tax. People leave money to the *falaj* in their wills (50 to 100 riyals) from time to time, and people with sufficient funds give when needed. Altogether these sources are sufficient for running the *falaj* and for maintenance. Some of the trees along the main *falaj* channel belong to the *falaj*, but do not provide a significant income. The *wakīl*s keep 2000 riyals in hand for emergency works and share 10% of the cost of major works, which comes to c. 150/yr each. In 2006, major repairs were carried out to 450 m of the tunnel section of the *falaj* costing 11,000 riyals, which was provided by the MWR.

[50] Salam bin Abdullah bin Dhiyab, 30 March 2007.
[51] 17 November 2006.
[52] The months of the Islamic calendar are listed in Appendix C.

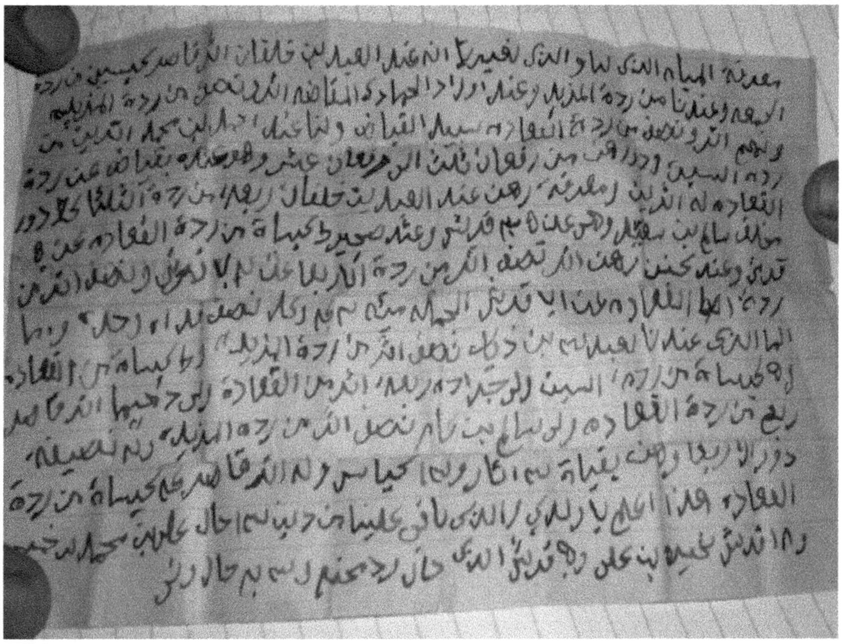

Figure 7.6 Document from Barzaman detailing water rights

Income		Expenditure		
Source	Amount, riyals	Purpose	Comment	Amount, riyals
auctions (*quᶜid*)	1840	time keepers		0
annual "tax" (*muzayyida*)	48	*wakīls*	c. 150 riyals each	300
falaj property	150	auctions		Little
trees along main channel	10	cleaning (*shiḥābī*) tunnel + channels	monthly: 50 riyals for tunnel, 10 riyals for surface channels	720 -1000
legacies	50 - 100	other maintenance		150 – 300
TOTAL	c. 2100		TOTAL	c. 1500

Table 7.2 Barzaman: summary of *falaj* finances c. 2006

7.3 METHOD OF STAR USE IN BARZAMAN

The stars are watched rising above a wall c. 3 m high (Figure 7.7). The observation position furthest from the wall is next to a small mosque on the opposite side of the road. There are markers of stones in the sand, in an east-west line between the mosque and the wall. These are not clear, as they are generally covered with silt (Figure 7.8), indicating that people know where to stand without the need for markers.

Starting next to the mosque and facing east, the observer watches a star rise above the wall and then moves east to the next marker nearer the wall to watch the same star rise at a later time, as illustrated in Figure 7.9. That is the theory, but to date I have only seen one person, a Hindu farm worker, clearly looking for their star, on 13 November 2006. It seems that those who use the star clock/dial are few, and come for a very short time just as a check. At my first meeting with Suleiman bin Saeed bin Saif (30 April 2006) he said that all old men in the village know the stars, but later that only four know them well and on cloudy nights (and days) wristwatches are used for timing the water shares.

Suleiman bin Saeed once said that he stays out all night and people come to ask him for information, but this is clearly not the case. Humaid bin Ali also said that he is at the star watching point every night, again not the case, although it is possible that either or both of them used to spend some time there each evening. From observation, it appears that most people watch their own star and go to Suleiman or Humaid in their houses for clarification. However, by October 2007, Humaid had practically retired from an active role in *falaj* timing, at least at night, saying that his eyesight was too poor to continue.

Figure 7.7 (left) **The stargazing point in Barzaman**

Figure 7.8 (below) **Looking for star markers in Barzaman**

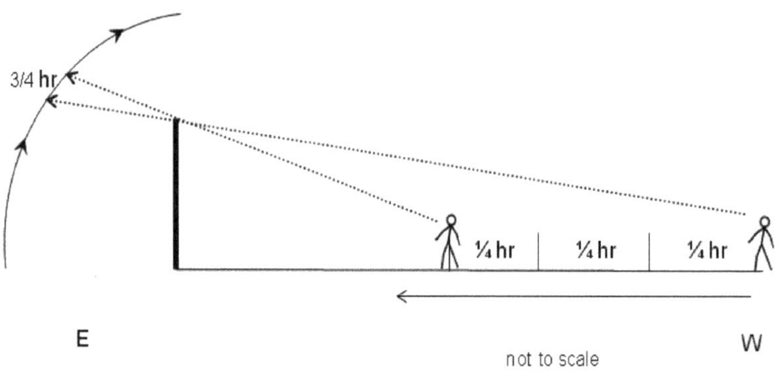

Figure 7.9 (left) **The method of stargazing in Barzaman**

Hamud bin Saeed bin Salim, who works in Ministry of Defence, is learning the star system from Humaid bin Ali. It seems that he is doing this for interest rather than as a career move, saying that it is difficult to use, both because of light pollution and because he is working. He said that the stargazing point was once further south, but was moved to be nearer the people now living in this area. The *wakīl* confirmed this: it was moved over 30 years ago, which is consistent with development of the village after 1970. The old stargazing point was also a wall 3 m high, at the southern edge of the area of irrigation (Figure 7.1). Hamud also said that some expatriate farm workers know the stars, which fits with my observations. Humaid has also been teaching his sons, so there is a group of younger people who could carry on the tradition.

7.4 STAR NAMES, IDENTIFICATION AND TIMINGS IN BARZAMAN

The names of stars used in time keeping[53] and their identifications are given in Table 7.3. As in other places there are several deviations from modern standard Arabic (MSA): one that is common in the study area is the shortening of the *alif* in Dharāᶜayn (or Dhirāᶜayn) to Dharaᶜayn. Uncertain identifications are given in italics.

The locations of the stars on the shortest and longest night of the year are shown in Figure 7.10. Each star has two *athar*s and the time between each star was given by Suleiman bin Saeed on 30 March 2007. However, each is said to be less than 60 minutes long and the total is only 40 *athar*s, short of the 48 *athar*s (c. 24 hours) needed for the full year: this information can, therefore, not readily be used as a check on the star identifications.

The second Ṣāra was also pointed out as Algenib (γ Pegasi) once, but this is too far from Scheat (β Pegasi), for which the identification is more certain. Capella (α Aurigae) is called both ᶜAyyuq ath-Thurayyā and Zāhira (shining). Thaqīla was said on one occasion to be two to four small stars rising south of Dharaᶜayn, but the identification as Denebola (β Leonis) was made more than once. Chuwī (Vega, α Lyrae) and its *ᶜayyuq*, Antares (α Scorpii), rise at roughly the same time, the top of the wall being more or less horizontal, so Antares cannot be used as a divider as it is in Qarya. The two stars of Mūfī rise six to seven minutes apart, but as with Hamal and Sheratan (α + β Arietis), this difference is not considered to be significant, at least in Barzaman.

[53] Mainly by Suleiman bin Saeed and Salam bin Abdullah on various occasions.

Star name as spoken		Comment (often related to location of star rise)	No. of stars	Mag-nitude	Classification*	English name*
Arabic	English					
الصارة الاولية	Ṣāra *l-awliyya*	once described as 2 or 3 stars in a line	1	2.43	β + *μ* Pegasi	Scheat, possibly with *Sadal Bari*
	ᶜayyuq	rises 22 min after Scheat close to al-Saᶜd, not bright	1	4.96	72 Pegasi	
الصارة الثانية	Ṣāra *th-thāniyya*		1	2.06	α Andromedae	Alpheratz
السعد	Saᶜd		1	2.06	β Andromedae	Mirach
الكوكبين	Kawkabayn		2	2.62 2.00	α + β Arietis	Hamal + Sheratan
الفتح	Fatḥ	bright, S of Kuwī, N of Kawkabayn + Thurayyā	1	2.06	*β Persei* or *41 Arietes*	*Algol*
الثريا	Thurayyā		cluster	>=2.84	Messier 45	Pleiades
	ᶜayyuq	26 min after	1	0.06	α Aurigae	Capella
بو قابل	Bū Qābil	Aldebaran is the *falaj* star	2	0.84 0.06	α Tauri α Aurigae	Aldebaran Capella
الشابك	Shābik	S of Mīyazīn (Orion's belt), described as a *khanjar*	cluster	≥ 2.75	cluster around Nair al Saif (ι Orionis)	Orion's sword
الظلمي	Ẓalmī	2nd star rises 13 minutes later, possibly the *ᶜayyuq*	1 or 2	1.90 3.34	γ Geminorum ξ Geminorum	Alhena Alzirr
الشعرة	Shaᶜra		1	0.37	α Canis Minoris	Procyon
الجنب	Janb	brightest in head of Hydra	3	3.37	δ Hydrae + 2	
الذرعين	Dharaᶜayn		2	1.34 2.00	α + γ Leonis	Regulus + Algeiba
لبطين	Buṭayn	2nd could be the *ᶜayyuq*	2	2.50	δ Leonis +1	Zosma
الثقيلة	Thaqīla	bright rising a little S of Buṭayn[3]	*1 or 4 small*	2.12	β Leonis	Denebola
الميثاب	Mīyathāb		2	2.84 3.43	ε + γ Virginis	Vindemiatrix + Porrima
الذكرين	Dhakarayn	N used for the *falaj*	2	-0.07 0.96	α Boötis + α Virginis	Arcturus (N)+ Spica (S)
الغفر	Ghafar (also Nisr)		1	2.21	α Coronae Borealis	Alphecca, Gemma
الزبان	Zubān	2nd star appears to be the *ᶜayyuq*	1 or 2 faint	3.71	γ Herculis	
	ᶜayyuq	5 minutes later	1	2.75	β Herculis	Kornephoros
القلب	Qalb		1	2.75	α Herculis	Rasalgethi
	ᶜayyuq	23 minutes later	1	2.06	α Ophiuchi	Rasalhague
الكوي	Chuwī		1	0.00	α Lyrae	Vega
	ᶜayyuq	rises at same time	1	1.03	α Scorpii	Antares
المنصف	Manṣaf	rises in E c. 50 min after and S of Chuwī	1 or 2	3.03 2.96	*β Cygni + ζ Aquilae*	*Albireo + Deneb el Okab*
الموفي	Mūfī		2	0.75 2.71	α Aquilae + γ Aquilae	Altair + Tarazed
لغراب	Ghurāb		4	3.62 – 4.40	α, β, δ +λ^2 Delphini	Sualocin, Rotanev (α, β)
اللادم	Ladam		1	2.37	ε Pegasus	Enif
**italics*: identification uncertain.						

Table 7.3 *Falaj* stars identified in Barzaman

A number of stars have an *ᶜayyuq*, showing that this is a description, not only part of the proper name for Capella of ᶜAyyūq al-Thurayyā, the name given in the literature. In Barzaman, *ᶜayyuq* is considered to mean *musāᶜid*, i.e."helping",[54] a positive connotation, and an indicator of the main star. It appears that dividers are not used.

One example of a saying related to agriculture was forthcoming, from Salam bin Abdullah (11 April 2008):

Lam tuṭlaᶜ ath-thurayyā min ᶜashayya: yazaraᶜ al-burr hayya hayya.

When Thurayyā rises in the evening, a lot of wheat is grown.

[54] Salam bin Abdullah and others in his family, 11 April 2008.

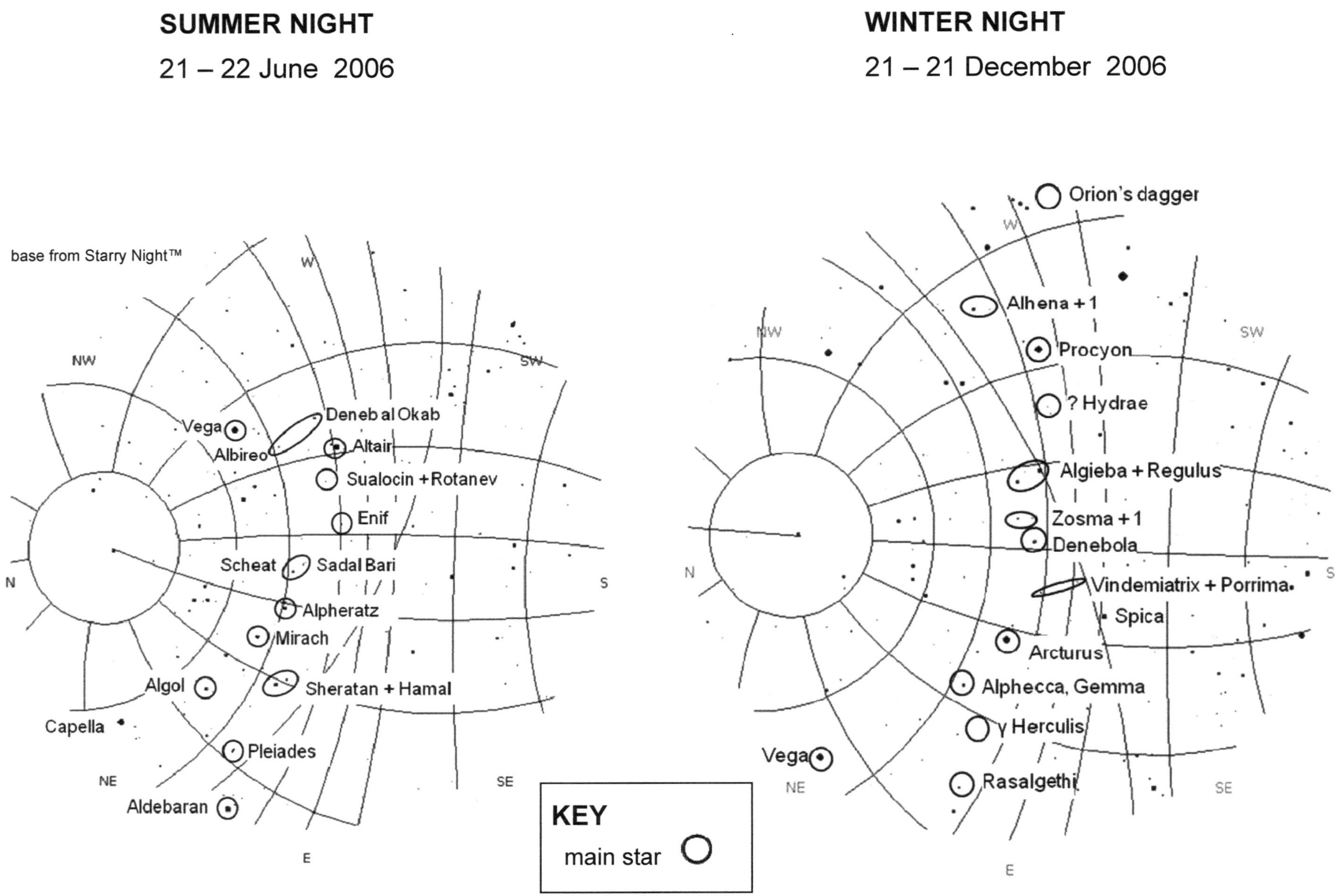

Figure 7.10 *Falaj* stars identified in Barzaman

8 SUDAYRA

8.1 INTRODUCTION

In Sudayra, there is one settled tribal group, the Hashmi, and at least thirteen Bedu tribes but the *falaj* was traditionally used only by the Hashmis.[55] The layout of the *falaj* is shown in Figure 8.1. No information was forthcoming on the history of the *falaj*. The *falaj* is still used for domestic purposes. The *shariᶜa* (traditional point of first use) is used for washing by men, followed by a mosque with wash rooms and a *mughsila* (wash room for corpses). There are several other small mosques with wash rooms along the main channel (Figure 8.2).

8.2 ORGANISATION OF THE *FALAJ* IN SUDAYRA

Hamad bin Shinun Al-Hashmi was both *wakīl* (administrator) and *ᶜarīf* (officer). He had been *wakīl* for some years and became *ᶜarīf* as well in 2004 when the former *ᶜarīf* died. There used to be auctions, but no longer, so there is no auctioneer. There are two heads of the Hashmis in the village but they do not get involved in running the *falaj* or settling disputes. Formerly there was a group of village elders who met from time to time to discuss *falaj* matters, but in recent years Hamad bin Shinun has, in effect, been on his own. Tired of problems with the Bedu not paying what was promised for cleaning and repairs to the *falaj* system, and carrying out works without consulting him, he resigned in 2007 since when the responsibility has passed to government offices in Mudaybi.

The irrigation cycle is seven days, and each cycle a person's allocation of water changes from day to night (Figure 8.3), in a similar fashion to Barzaman. The time at night is divided into *athar*s, while that by day is in *sahm*s (1½ *athar*s or, nominally, ¾ hour), although, according to Hamad, some people use the term *athar* for the day as well. He himself sometimes uses the term *sahm* when referring to the night. The *badda*s, which are of ½-day duration, did have names some 200 years old, but they are no longer used and he has forgotten most of them, except *bishayra*, possibly meaning something related to "good tidings" and *suwaylū* possibly from *suwayl*, an equal (Lane, 1968. V4: 1474).

The use of stars and sun is thought to go back as long as the *falaj*.[13] It appears that the sundial (Figure 8.4) is still used by many, but the stars by only a few. Hamad thinks he got his first watch in the early 1970s when they became more available, but it appears that it was the development of the new village and abandonment of the old that precipitated the move away from traditional practices rather than simply the availability of watches.

Siyada bint Salim bin Mahmud Al-Yahafi has the largest share of water from the *falaj*, of twenty-four *athar*s, followed by Salam bin Saleem bin Khamis Al-Hindi Al-Wuhaybi with fifteen *athar*s, both from Bedu families. Hamad bin Shinun himself has six, and the majority of farmers have holdings of five *athar*s.

In each seven-day cycle, three *athar*s are used to raise money for the *falaj* service. This water was formerly auctioned annually, raising approximately 50 riyals each time, but now it is not auctioned, but sold at the fixed rate of 50 riyals. The *falaj* also owns 120 *athar*s, more than any individual, and worth approximately 2000 riyals/year.[56] Nowadays there is no *waqf* (mosque property) holding of *falaj* water as the mosques and schools etc. are supported by the government. In 2006, approximately 70,000 riyals was received for major repairs from two sheikhs in Abu Dhabi owning date gardens and some *falaj* water. It appears that at least one of the sheikhs was originally from Sudayra. In 2005, Hamad wrote a summary of the money raised for the *falaj* covering the last 100 years in blocks of 16 *sahm*s (equivalent to 24 *athar*s or c. 12 hours). I was shown this briefly on 15 November 2006, but was asked not to photograph it as the information is confidential to the people named therein. Information on *falaj* expenses was not provided, for the same reason.

8.3 METHOD OF STAR USE IN SUDAYRA

The official stargazing point is the outside wall of the old village (Figure 8.4), in the same location as the sundial. The old village was abandoned and fell into ruins about fifteen years ago (i.e. c. 1992). Before this, the stargazing point was the wall of a mosque inside the ruins. When the village was abandoned and the stargazing point changed to the current position, many people stopped using the stars. Although on my initial visit (23 January 2006) it was said that many people use the stars, later the estimate fell to ten or so and I was advised not to spend any time at the location waiting for someone to come, as it was unlikely that anyone would. Apparently the few who still use the stars watch from their houses or gardens.

[55] Hamad bin Shinun Al-Hashmi, 25 March 2007.

[56] Hamad Shinun, 14 November 2006.

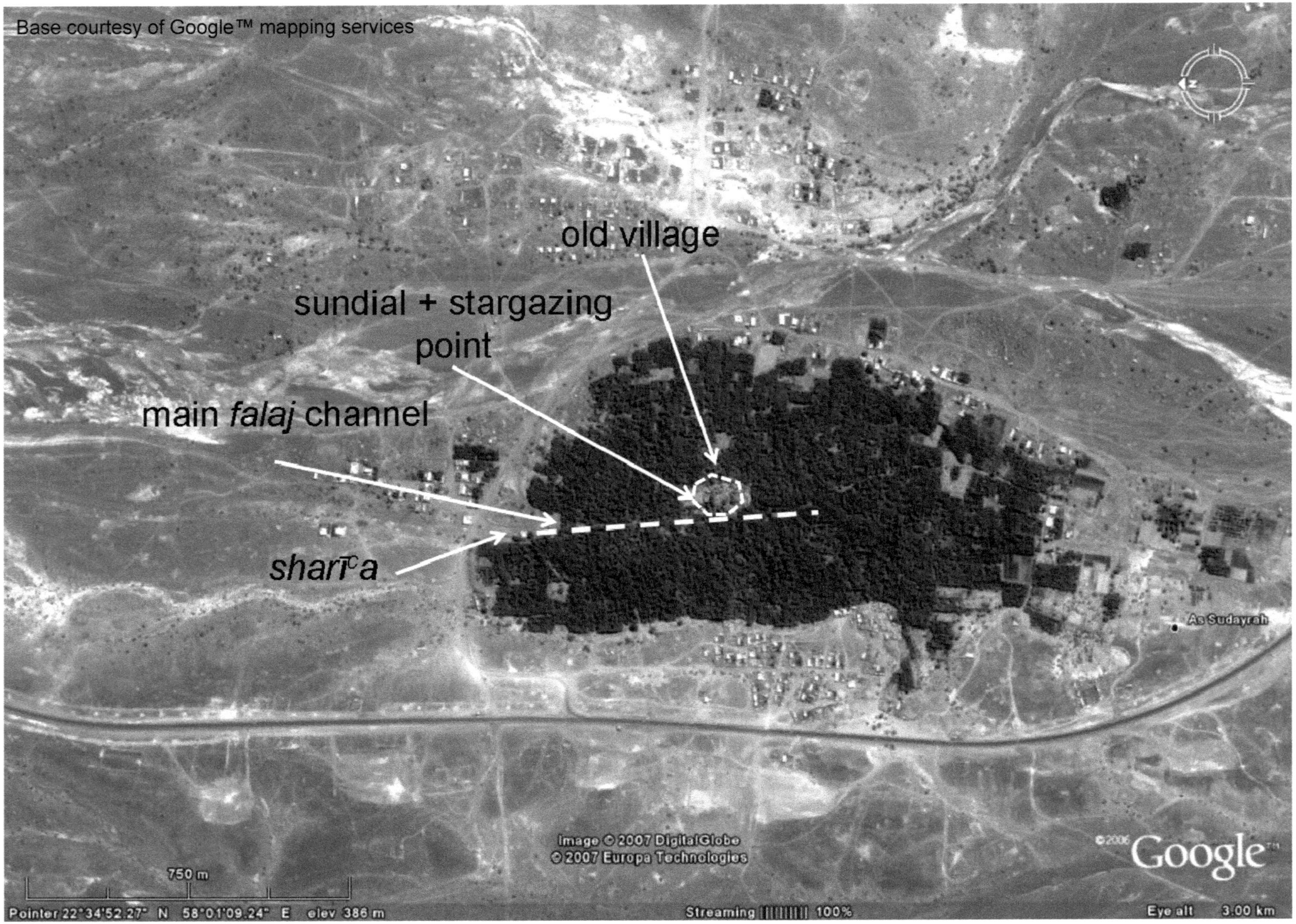

Figure 8.1 Layout of Sudayra

Figure 8.2 Wash rooms on the *falaj* in Sudayra

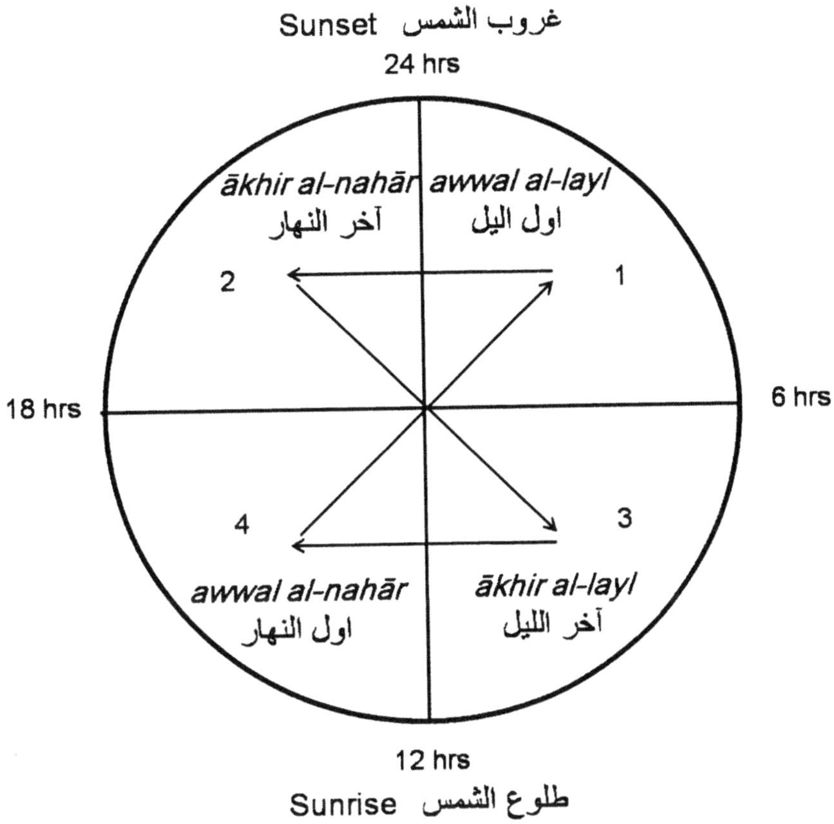

Figure 8.3 Rotation of water shares in Sudayra

Figure 8.4 Stargazing point and sundial in Sudayra

There are no special marks on the wall. Facing east, people watched stars rise above the palm trees, using spaces along the wall marked only by the number of steps. Only Iklīl (see section below on star names) was not watched from the wall, but from a palm tree to the north of the wall, from where it is possible to see further south.

8.4 STAR NAMES IN SUDAYRA

Hamad bin Shinun Al-Hashmi is credited with being the most knowledgeable and was the only person to provide information on the stars in Sudayra, pointing them out in the sky and using Starry Night™. It is clear that different stars are used when the main star is not visible, but some variants were too far out time wise to be feasible and have therefore been ignored. The star names as spoken, given by Hamad bin Shinun on different occasions, and identifications are given in Table 8.1 below, and their locations in the sky are shown in Figure 8.5.

Although γ Cassiopeiae was pointed out as the second star of the first Ṣāra, it could alternatively be ε Cassiopeiae, which rises at the same time as Schedar (α Cassiopeiae). If the stars of Shābik cannot be seen, Bellatrix (γ Orionis) and Orion's belt (Mintaka, Alnilam, Alnitak – δ, ε, ζ Orionis) are used as indicators of the main star. Since the cluster of Thaqīla is quite faint (the brightest star has a magnitude of 4.71) it is likely that everyone now uses Cor Caroli (α Canum Venaticorum), which rises a little beforehand and is much brighter, at a magnitude of 2.84.

The list contains two star pairs: Dhakarayn (Arcturus and Spica), in which both stars are used, and Fatḥ (Algol and Mirfak), in which either is used.

The use of stars in Sudayra is now infrequent; it seems that the stargazing wall is used so little that people are forgetting the method. Sudayra is not typical of the other villages included in the study: the settling of Bedu with no tradition of communal *falaj* management, and the financial dependence on a few wealthy people such that there is no need for the system to be self-supporting financially are clearly major factors in the recent decline of traditional practices. The development of the new village in c. 1992, although later than in other places, was also a major factor in the change.

8.5 OTHER TALES ABOUT STARS IN SUDAYRA

No information was collected on agricultural sayings related to stars in Sudayra. However, giving a lift to a Bedu woman, Salima bint Kharaybish Al-Musalimi, who was hitchhiking home to Sudayra from the hospital in Ibra on 1 May 2006, resulted in an invitation to coffee, and gave me the opportunity on my next visit, in November 2006, to ask about the use of stars by the Bedu for navigation. I was told that they do not use the stars anymore, as there are now roads. Salima and her father told me how the Arabs of old, i.e. before the days of Islam, buried their dead with the head facing Yēḥ, the Pole Star, and that the rise of Kuwī is associated with very cold weather, so cold that animals die. Thurayyā and Shābik were also mentioned as stars that move in the sky, unlike Polaris.

Star name*		Comment	No. of stars	Magnitude	Classification**	English name
Arabic	English					
الصارة	Ṣāra		2/3	2.21	α + *γ +/ε* Cassiopeiae	Schedar (α)
الصارة الخلافية	Ṣāra (*al-khalāfiyya*)		1	2.06	α Andromedae	Alpheratz
السعد	Saᶜd		1	2.06	β Andromedae	Mirach
الكوكبين	Kawkabayn		2	2.62 2.00	α + β Arietis	Hamal + Sheratan
الفتح	Fatḥ	either star used, depending on visibility. Each is called al-Fatḥ	1	2.06 1.78	β Persei α Persei	Algol Mirfak
الثريا	Thurayyā		many	>=2.84	Messier 45	Pleiades
المقبل	Maqbil/Muqbil	smaller than Kuwī, red yellow	1	0.84	α Tauri	Aldebaran
الشابك	Shābik		3	3.37 – 4.37	λ, $\varphi^1 + \varphi^2$ Orionis	Meissa (λ)
الظلمي	Ẓalmī		1	1.90	γ Geminorum	Alhena
الشعرة	Shaᶜra		1	0.37	α Canis Minoris	Procyon
الجنب	Janb	use brightest of top group	1	3.37	ε Hydrae	
الذراعين	Dharāᶜayn		2	1.34 - 2.00	α + γ Leonis	Regulus + Algieba
البطين	Buṭayn/Boṭayn		1	2.50	δ Leonis	Zosma
الثقيلة	Thaqīla	faint stars S of and between 6th and 7th stars in Plough; when this cannot be seen, use Cor Caroli	cluster	≥4.71 2.84	around 20 Canum Venaticorum; α Canum Venaticorum	 Cor Caroli
الميثاب	Mīyathāb		2	2.84 3.43	ε + γ Virginis	Vindemiatrix + Porrima
الذكرين	Dhakarayn	use both stars	2	-0.07 0.96	α Boötis + α Virginis	Arcturus + Spica
الغفر	Ghafar		1	2.21	α Coronae Borealis	Alphecca, Gemma
الزبان	Zubān		1	2.75	β Herculis	Kornephoros
الاكليل	Iklīl		2	2.53 2.28	β + δ Scorpii	Graffias + Dschubba
الكوي	Kuwī		1	0.00	α Lyrae	Vega
المنصف	Munṣif/Munṣaf		1	2.96	ζ Aquilae	Deneb el Okab
الموفي	Mūfī	also called *an-nisr* (eagle)	1	0.75	α Aquilae	Altair
لغراب***	Ghurāb	described as a bird	4	3.62 - 4.40	α, β, δ + λ^2 Delphini	Sualocin (α), Rotanev (β)
اللدم	Ladam		1	2.37	ε Pegasi	Enif
* as spoken by Hamad Al-Hashmi; ** *italics*: identification uncertain ; *** pronounced "la-Ghurāb" not "al-Ghurāb".						

Table 8.1 ***Falaj*** **stars identified in Sudayra**

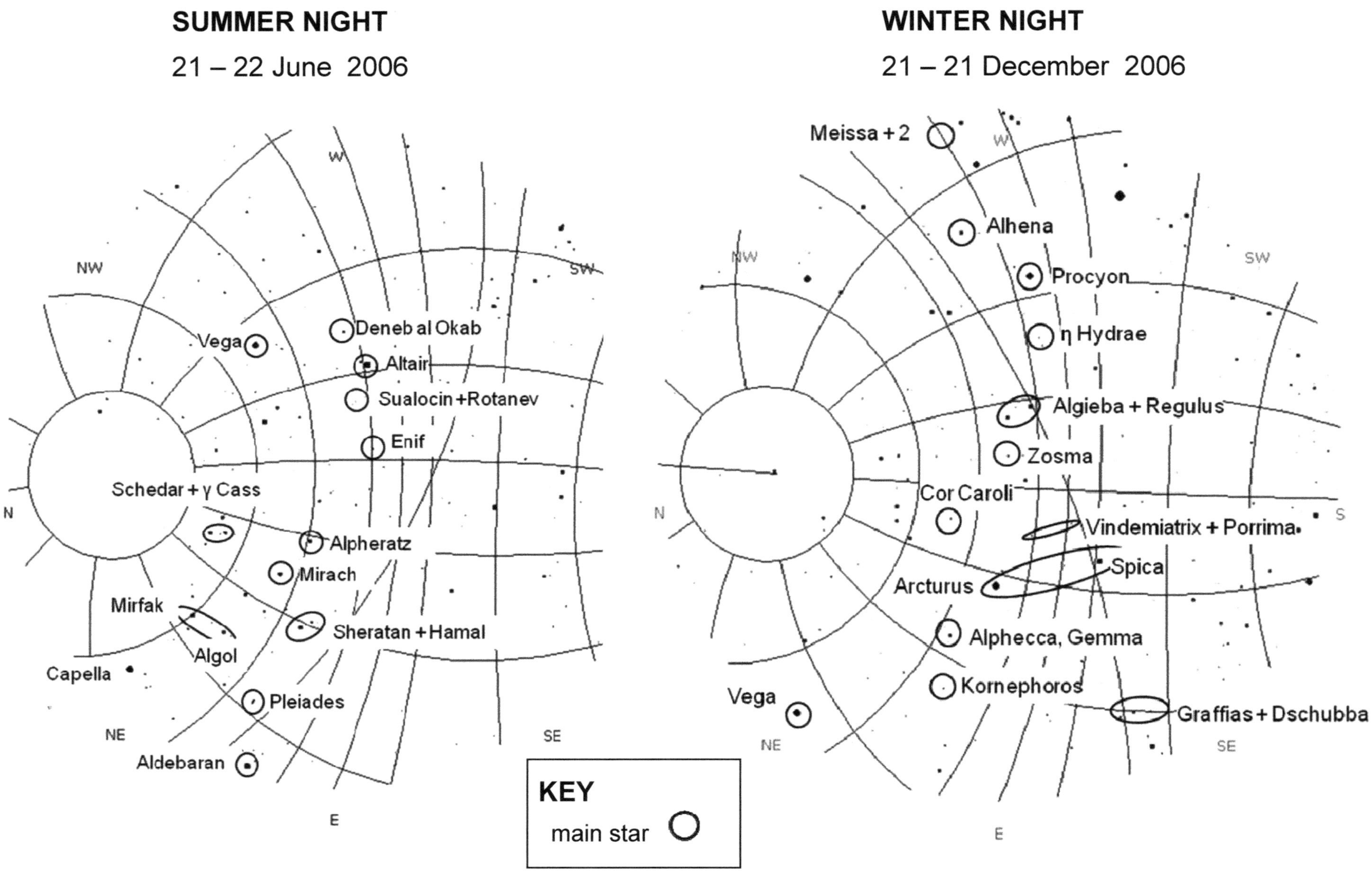

Figure 8.5 *Falaj* stars identified in Sudayra

9 AL FATH

9.1 INTRODUCTION

Al Fath is a town with at least ten tribal groups. The *falaj* has been present as long as the town, possibly for 350 years.[57] The layout of the town and *falaj* are shown in Figure 9.1. The old town is still lived in, with the newer development around. The *falaj* flow has decreased significantly in recent years (see section 5.2): according to Sultan Al-Shidhani, my main source of information, in 2002 one *athar* (c. ½ hr) was sufficient to irrigate 30 trees, but by 2006 was only sufficient for five to ten trees.

The *falaj* channels are used for washing, with bath houses still in use (Figure 9.2). As in other places, the *falaj* is used near the mosques for ablutions before prayers. There is also at least one building for washing corpses. The *sharīʿa* (former point of first use for drinking water) is located just outside the north gate of the old town and is no longer used, but is well protected and kept in good condition. Between the *sharīʿa* and the stargazing point, the channel flows through the compounds of private houses, where doubtless the water is put to a variety of uses.

9.2 ORGANISATION OF THE *FALAJ* IN AL FATH

The *falaj* officials are listed in Table 9.1. It seems that most of the work is done by the *wakīl* (administrator), whose father was *wakīl* before him, since the *ʿarīf* (officer) is often out of town.

Title	Name	Role	Started	Other work
wakīl	Saud bin Humaid Al-Habsi	deals with finances	c. 1994	teacher
ʿarīf	not given			business-man
dillāl	Hamdun bin Humaid Al-Habsi	auctioneer		n.k.

Table 9.1 Al Fath *falaj* officials

Decisions on *falaj* matters, including the appointment of officials are taken by a council, of which Sheikh Sultan bin Mohammed Al-Habsi, the head of the Habsis in the Mudaybi area, is a member. Unlike Sudayra there are no significant disputes; certainly nothing that cannot be dealt with by the *wakīl*. Sometimes people take more than their fair share, but this is usually by mistake and can readily be rectified the next time around.

Information on the *falaj* cycle and water shares was provided by Sultan Al-Shidhani on different occasions, and by Saud bin Humaid Al-Habsi, the *wakīl* of the *falaj*.[58] The irrigation cycle is said to be eighteen days. Each 24-hour day is one *bāda* (*badda*), which is divided into two *farda*s; the end and beginning of the day are treated as one long *athar*; it appears that the term *sahm* is not used. The stars and sun are thought to have been used as long as the *falaj* has been in existence; the water clock is not known. The names of the main blocks of water are given in Table 9.2. Several are named after tribal groups (eg *mā' awlād mahra* – "water of the sons of *mahra*"), others (the two *qaʿds*) for the *falaj* share.

The total time shown in Table 9.2 is 18½ days, so there is possibly an error either in this list or the reported length of the irrigation cycle. The allocation of water changes from day to night every nine days and is at the same time after sunset at night as after sunrise during the day (Figure 9.3), as in Zahib. The owner of the largest share of water is Sheikh Sultan, with 27 *athar*s. Sultan al-Shidhani does not own any water rights but buys what he wants at auction, as, he says, do many others.

The location of the sundial is shown in Figure 9.4. It is a few minutes walk from the stargazing point. The divisions are in *athar*s - 12 each side of the pole, and there are numerous personal time markers, indicative of frequent use.

Saud bin Humaid Al-Habsi said that it was too complicated to provide a summary of the *falaj* finances for one year. The income from auctioning two days of water each irrigation cycle is 1000 riyals/year, but some people work in the United Arab Emirates and give larger amounts for major repairs, providing 15,000 to 50,000 riyals in any one year. In 2007 it cost 1000 riyals to buy one *athar* outright. Records of sales are kept by the *wakīl* and he informs the government offices; apparently people do not keep their own records. Each person does their own time keeping and the *falaj* officials are not paid. There are no taxes on water or land use or the crops produced, and the *falaj* has no *waqf* (mosque) property.

Approximately 1.5 km of the *falaj* tunnel section was repaired in c. 2007 with government funds, and an extension, needed because of declining flows, is being built with *falaj* funds.

[57] Sheikh Sultan bin Mohammed Al-Habsi, 20 October 2007.

[58] 25 March 2007.

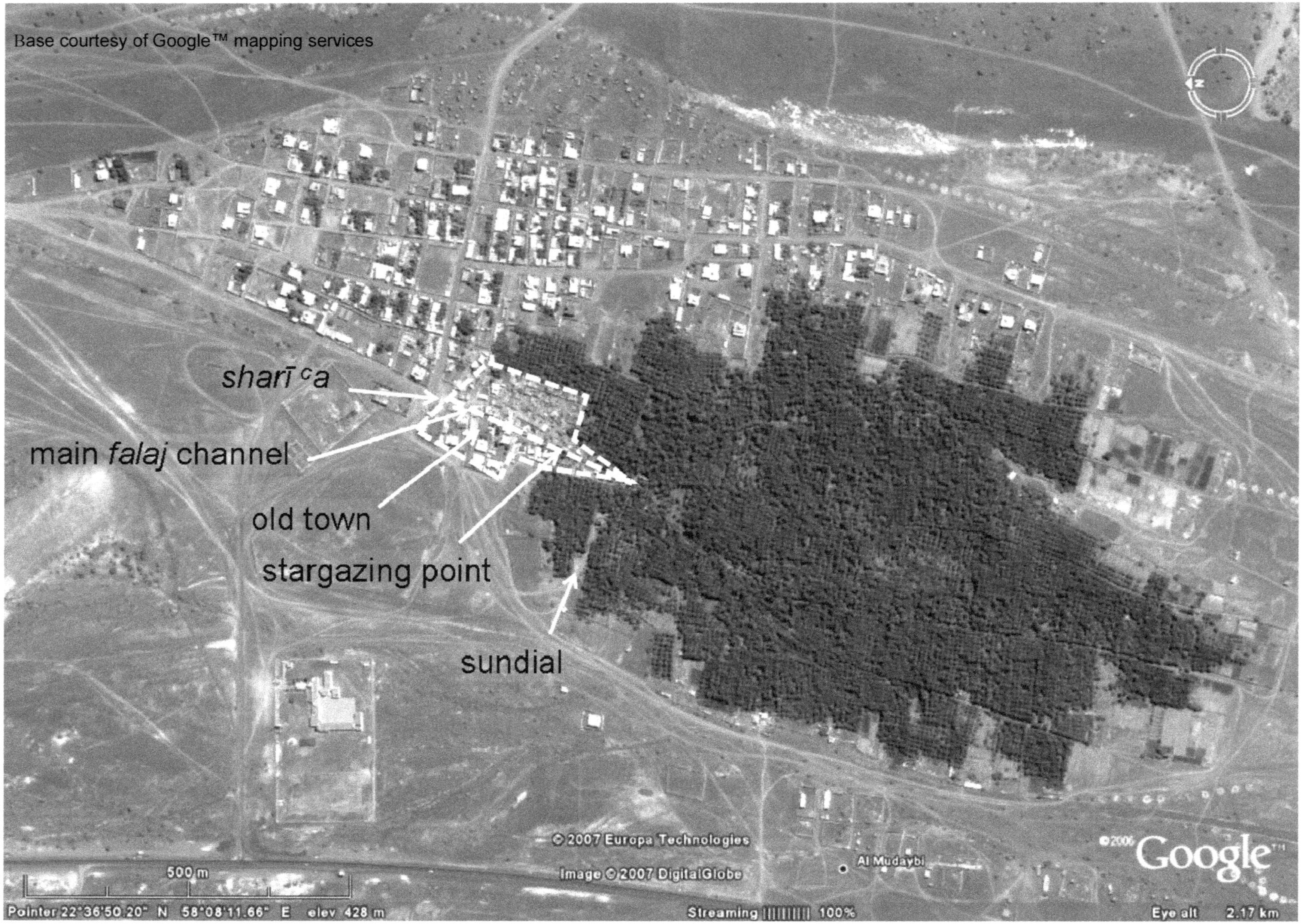

Figure 9.1 Layout of Al Fath

Figure 9.2 Public baths and lit access to the *falaj* in Al Fath
1 April 2007

Major water shares in Al Fath		
Name as written	**Transliteration**	**Hours duration**
ماء عامر	*mā' ^{c}āmar*	48
ماء البراونة	*mā' al-b.rawna*	48
البادة التي تلي ماء البراونة	*al-bāda allati talī mā' al-b.rawna*	24
ماء العاسوة	*mā' al-^{c}āsiwa*	48
الفردة الميتاعة	*al-farda l-maytā^{c}a*	12
بدوت البادة	*badūt al-bāda*	24
قعد	*qa^{c}d*	24
ماء الخدم	*mā' al-khadam*	48
ماء اولاد مهرة	*mā' awlād mahra*	48
ماء الشراك	*mā' al-sharāk*	48
فردة خميس	*farda khamīs*	12
بدوت البادة	*badūt al-bāda*	24
قعد	*qa^{c}d*	24
الفردة التي يقعد الربع	*al-farda allatī yaqa^{c}d al-rub^{c}*	12

Table 9.2 Major water shares in Al Fath

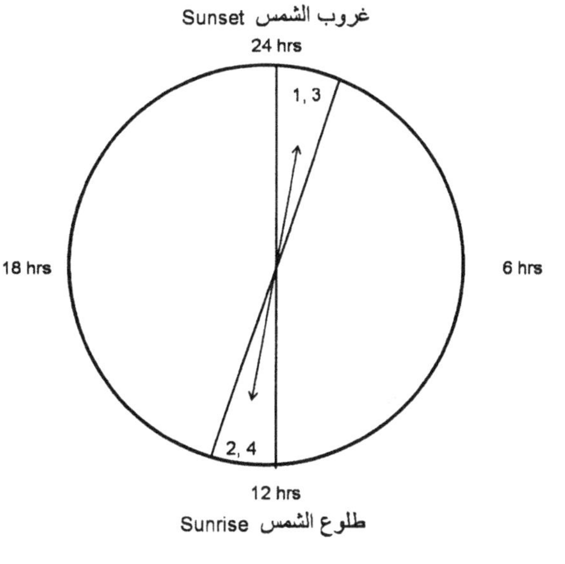

Figure 9.3 Rotation of water shares in Al Fath

Figure 9.4 The sundial at Al Fath
Ali bin Aamar Al-Habsi MRW, 11 November 2006

9.3 *FALAJ* STARS AND METHOD OF WATCHING STARS IN AL FATH

Table 9.3 gives the local names of the 24 main stars identified in Al Fath with their English names and classification, and the time that they are said to represent. The space between each main star is two *athar*s. While the times do not add up to 24 hours, some stars have more and some less than one hour, and the times must therefore be closer to reality than those for Barzaman, which were all said to be less than one hour. The only stars mentioned as weather or agricultural indicators were Saᶜd, when there is no cultivation,[59] Dhalmī, a time of great heat, and ᶜAqrab, which brings rain.[60]

The two Ṣāras are groups of three relatively faint stars showing almost the same pattern, different to the brighter stars known as Ṣāra in other villages. The first Ṣāra lies on the celestial equator, with about 10 minutes between the first and last star. The second lies at approximately 10°N and has 20 minutes between the first and last star, i.e. the stars are more spread out than in the first. Shābik is described by Sultan Al-Shidhani as a fishing net and seems to comprise much of the first half of Orion, as seen rising, but only the one star, Bellatrix (γ Orionis), is used for time keeping. These two stars are illustrated below.

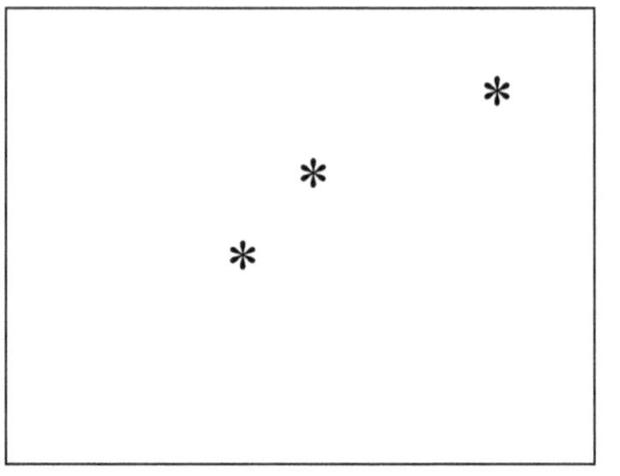

Ṣāra in Al Fath

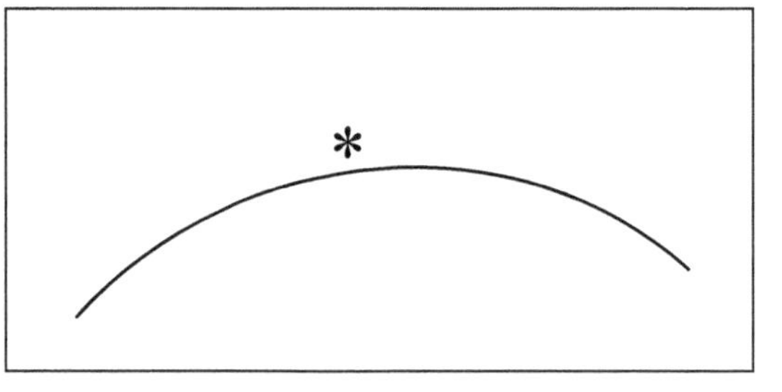

Shābik in Al Fath

The method of star use does not involve looking at the horizon: the main stars are not watched directly, but other stars are used as indicators of the time of rise and divisions between the main stars. The indicator stars are watched rising, setting or in the zenith relative to various markers described later in this chapter. The stars watched to know the timings of the twenty-four main stars are called *maqiyās*, meaning "for measurement". Only five such stars are used:

Dhakarayn,

Kuwī,

ᶜAyyāq ath-Thurayyā,

Thurayyā, and

Shaᶜarā.

These are all bright and readily recognisable stars. Dhakarayn. Kuwī and Thurayyā are also main stars and are listed in Table 9.3. ᶜAyyāq ath-Thurayyā is Capella (α Aurigae) and is also known here as Zāhara; the Shaᶜarā watched is Sirius (α Canis Majoris), not the main *falaj* star, Procyon (α Canis Minoris). The locations of the main stars and the *maqiyās* (indicator stars) in the sky are shown in Figure 9.5. It is said that everyone uses the stars except when it is cloudy, when wristwatches are used.

[59] Saud bin Humaid al-Habsi, *wakīl*, 25 March 2007.
[60] Sultan Al-Shidhani, 17 and 20 October 2007.

Star names*		No. of stars*	Minutes to next star*	Magnitude	Classification	English name
Arabic	English					
الصارة الولى	Ṣāra l-Ūlā	3	45	3.84 – 4.28	γ, ζ1 + η Aquarii	Sadachbia (γ)
الصارة الثانية	Ṣāra th-Thāniyya	3	45	3.68 – 4.25	γ, θ + ι Piscium	
السعد	Saᶜd	1	50	2.06	β Andromedae	Mirach
الكبكبين	Kabkabayn	2	40	2.62 2.00	α + β Arietis	Hamal + Sheratan
الفتح	Fatḥ	1	60	3.59	41 Arietis	
الثريا	Thurayyā	cluster	60	>=2.84	Pleiades	Pleiades
الدبران	Dubrān	1	60	0.84	α Tauri	Aldebaran
الشابك	Shābik	1	70	1.62	γ Orionis	Bellatrix
الذلمي	Dhalmī	1	75	1.90	γ Geminorum	Alhena
الشعرى	Shaᶜarā	1	60	0.37	α Canis Minoris	Procyon
الجنب	Janb	1	60	3.09 3.37	ζ or ε Hydrae – the brightest	
الذراعين	Dharāᶜayn	2	60	1.34 2.00	α + γ Leonis	Regulus + Algeiba
البطين	Boṭayn	1	60	2.50	δ Leonis	Zosma
الثقيلة	Thaqīla	1	45	2.12	β Leonis	Denebola
الميثاب	Mīyathāb	1	60	2.84	ε Virginis	Vindemiatrix
الذكرين	Dhakarayn	1	60	-0.07	α Boötis	Arcturus
الغفر	Ghafar	1	60	2.21	α Coronae Borealis	Alphecca, Gemma
الزبان	Zibān	1	50	2.75	β Herculis	Kornephoros
الاكليل	Iklīl	1	40	2.75	α Herculis	Rasalgethi
لكوي	Kuwī	1	60	0.00	α Lyrae	Vega
المنصف	Manṣaf	1	60	3.03	β Cygnus	Albireo
النسر	Nisar	2	50	0.75 2.71	α + γ Aquilae	Altair + Tarazed
لغراب	Ghurāb	4	60	3.62 - 4.40	α, β, δ + λ^2 Delphini	Sualocin (α), Rotanev (β)
الادم	Adam	1	60	2.37	ε Pegasi	Enif
* information provided orally by Sultan Al-Shidhani.						

Table 9.3 ***Falaj*** **stars identified in Al Fath**

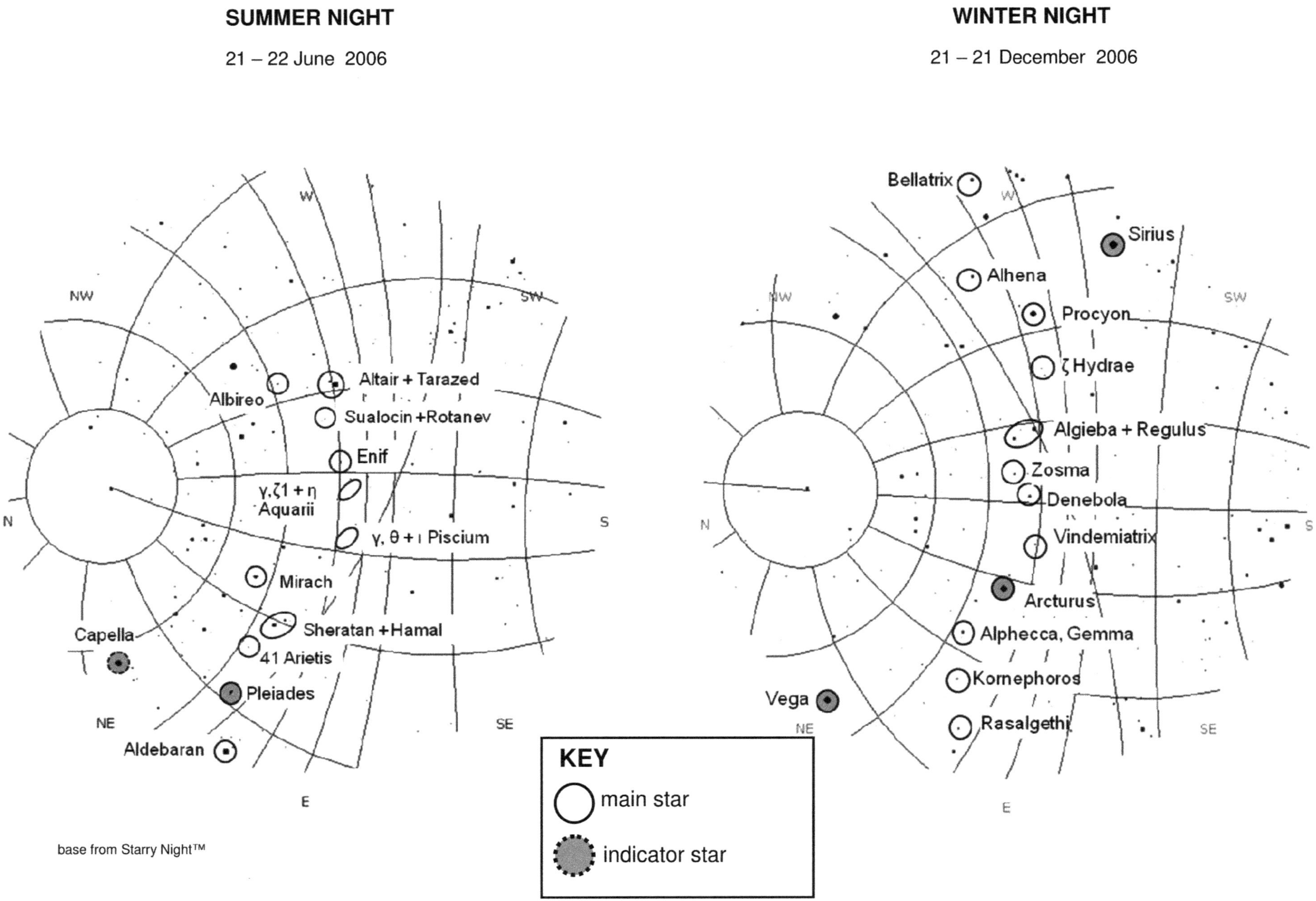

Figure 9.5 *Falaj* stars identified in Al Fath

There are many points for watching from, both inside and outside the mosque. A sketch plan of the area is given in Figure 9.6. Each line or point that is watched from is given a letter to facilitate the description of the method: this letter, the main star, the star used to mark it, the location watched from and direction of watching are listed in Table 9.4. It is not only the rising of the main star that is indicated by wall markings, but also intermediate points for dividing the space between two main stars into single *athar*s. This may become clearer when reading the description and seeing the further illustrations below. As mentioned previously, the stars rise four minutes earlier each night. Hence, every eight days or so, a different main star or *athar* marker between main stars is used. Personal markers of stones, cloth or paper are placed on top of the walls or fixed to nails in the walls to mark individual times.

Wall A is the north wall of the inner compound of the mosque. Dhakarayn (Arcturus, α Boötis) is the indicator for Ghurāb, Adam, the two Ṣāras and Saᶜd. As illustrated in Figure 9.7, the observer looks west, watching stars setting behind the wall. He starts at the wall and then moves back, eastwards, so that the star is seen setting at a later time.

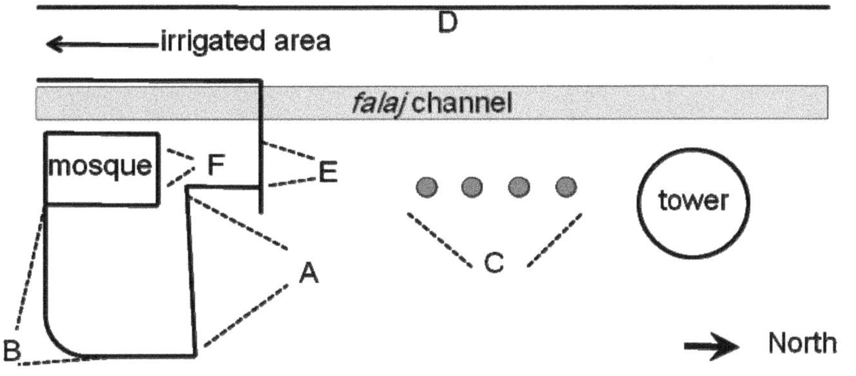

Figure 9.6 Sketch of the stargazing area in Al Fath
not to scale

Main star (from Table 9.3)	Star watched	Point watched from	Area on Fig. 9.6	Direction of view
Ṣāra l-Ūlā	Dhakarayn (setting)	N wall of inner compound of mosque	A	W
Ṣāra th-Thāniyya				
Saᶜd				
Kabkabayn	Kuwī (setting)	S wall of inner compound of mosque	B	W
Fatḥ				
Thurayyā				
Dubrān				
Shābik				
Dhalmī 1				
Dhalmī 2	ᶜAyyāq ath-Thurayyā (rising)	outside mosque	C	NNE to top of tower
Shaᶜarā				
Janb				
Dharāᶜayn 1				
Dharāᶜayn 2	Thurayyā (setting)	W wall outside mosque	D	W and upwards
Boṭayn 1				
Boṭayn 2	Shaᶜarā (Sirius) (rising)	N wall, outer compound of mosque	E	SE
Thaqīla				
Mīyathāb 1				
Mīyathāb 2	Shaᶜarā (Sirius) (setting)	nail inside N arch of mosque	A	SW corner of mosque
Dhakarayn		N wall of inner compound of mosque		
Ghafar				
Zibān				
Iklīl				
Kuwī	Dhakarayn (rising)	S wall, outer compound of mosque	F	E
Manṣaf				
Nisar				
Ghurāb	Dhakarayn (setting)	N wall, inner compound of mosque	A	W
Adam				
Where the main star is given twice, followed by 1 or 2, 1 refers to the first and 2 to the second *athar*.				

Table 9.4 Stars watched in Al Fath

The points from which Dhakarayn is watched, and the stars that it applies to, are shown in a composite picture of the wall in Figure 9.8. The observer starts at a nail in the wall of the arch and moves backwards to the furthest marker. From here, he will move to areas B, C, D and E before returning six months later to watch Sha^carā (Sirius, α Canis Majoris) setting from the upper line of markers. Sirius is watched setting not directly to the west as for Dhakarayn, but to the south west behind the mosque.

In practice the observer will be more than one person: each farmer, or their assistant, will normally only come to the appointed place for a few minutes to check their time before going to the fields to water their land.

Wall B is the S and E wall of the inner compound of the mosque, Vega (Kuwī, α Lyrae) is watched setting behind the SW corner of the mosque for several stars, starting with Kabkabayn. The points for observing the different stars are shown in Figure 9.9: the last point on this wall, which is off the picture to the left, is for the start of Alhena (Dhalmī). On top of the wall are a few stones put there to mark certain times by individual shareholders. This indicates active use of the system of stargazing and helps to confirm the assertion that it is used by everyone.

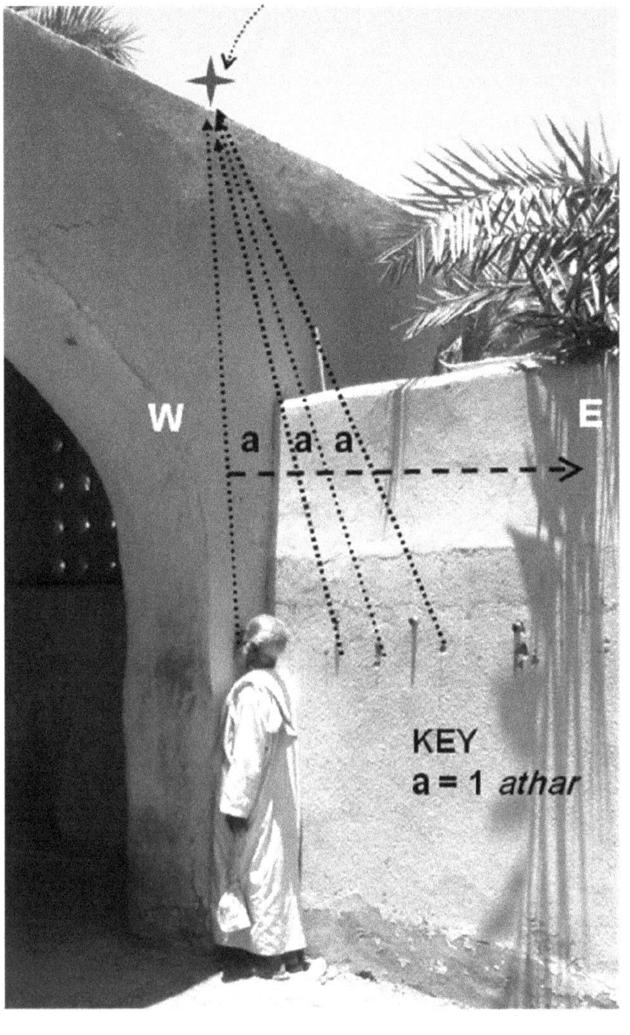

Figure 9.7 Method of watching Dhakarayn (Arcturus) setting in Al Fath
demonstrated by Mohammed bin Aamar Al-Habsi
30 April 2006

Line C is a line of six stones outside the mosque, leading up to the tower, shown schematically in Figure 9.10, with the line of sight in Figure 9.11. The first stone marks the start of the second half of Dhalmī and the last is the first point for Dharā^cayn. Capella (^cAyyāq ath-Thurayyā or Zāhara, α Aurigae) is watched rising above the top of the tower, and because this is curved, the stones also form a shallow curve.

Point D is on a wall to the west of the *falaj* channel (Figure 9.12). From here, Thurayyā (the Pleiades) is watched, standing against and facing the wall (i.e. facing west) looking upwards. The first marker is when the observer is standing up against the wall and the Pleiades reaches the top of the wall, i.e. it is in the zenith. This is the start of the second *athar* of Dharā^cayn. The second point is when the observer stands two fist widths away from the wall and the Pleiades meets the top of the wall. This marks the first *athar* of Boṭayn.

Wall E is the N wall of the outer compound of the mosque. Turning left inside the entrance gate to the mosque there are several vertical scratches and nails from which Sha^carā (Sirius) is watched rising in the south east above the mosque wall. The line of sight and the star markers are shown in Figure 9.13. From the last marker, closest to the wall, the next observation point is a nail inside the N arch of the mosque: the start of the sequence of stars for which Sirius is watched setting from wall A.

Wall F is S wall of the outer compound of the mosque, opposite the gate to the compound. Looking east from the observation points, Dhakarayn is watched rising, marking the times for Kuwī, Manṣaf and Nisar. The line of sight and star markers are shown with Wall E in Figure 9.13. Following this the sequence stars again using Wall A to watch Dhakarayn setting.

This system of stargazing is clearly more complex than any yet described in this study, but has the advantage that relatively few stars need to be known, and they are stars that are easy to see and to recognise.

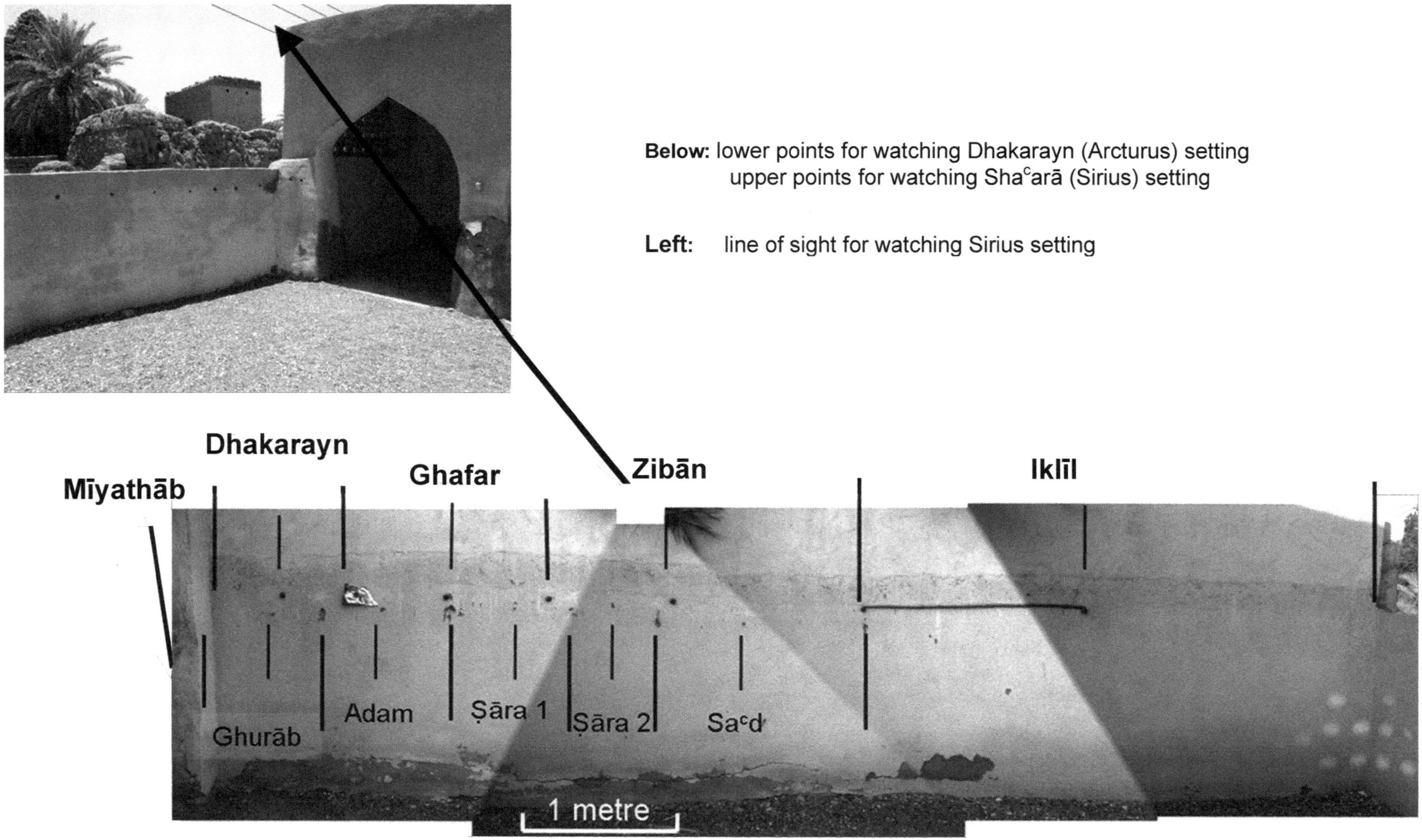

Figure 9.8 Star markers in Al Fath: wall A

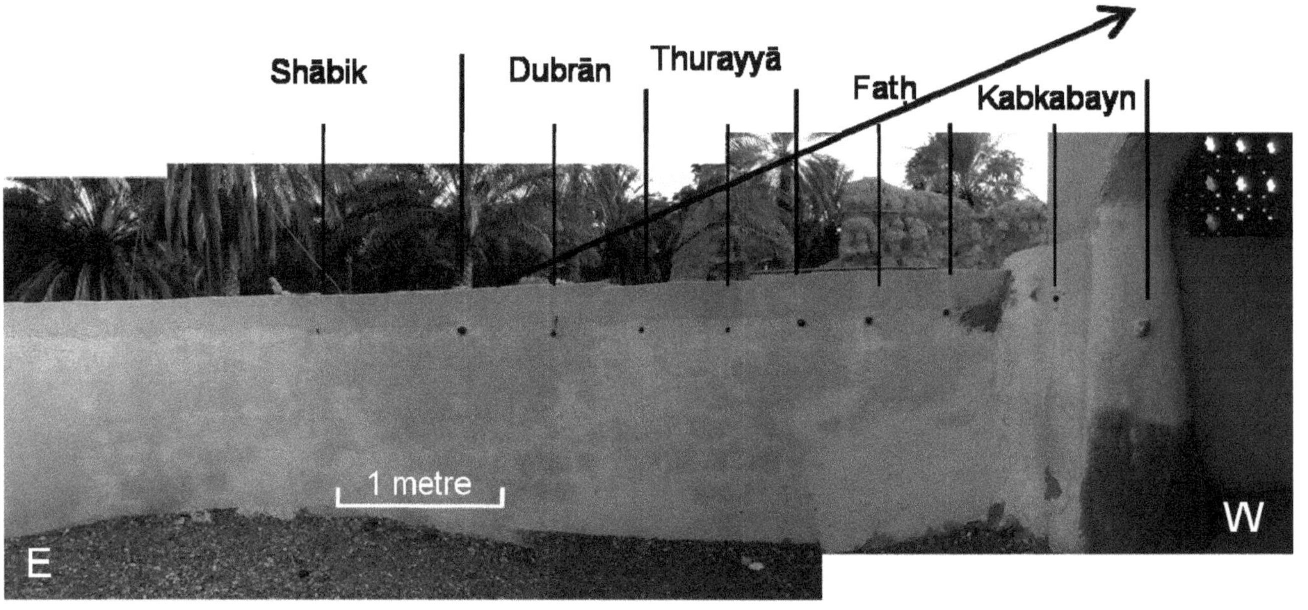

Figure 9.9 Star markers in Al Fath: wall B
with line of sight for Kuwī setting

Figure 9.10 Star markers in Al Fath: line C
not to scale

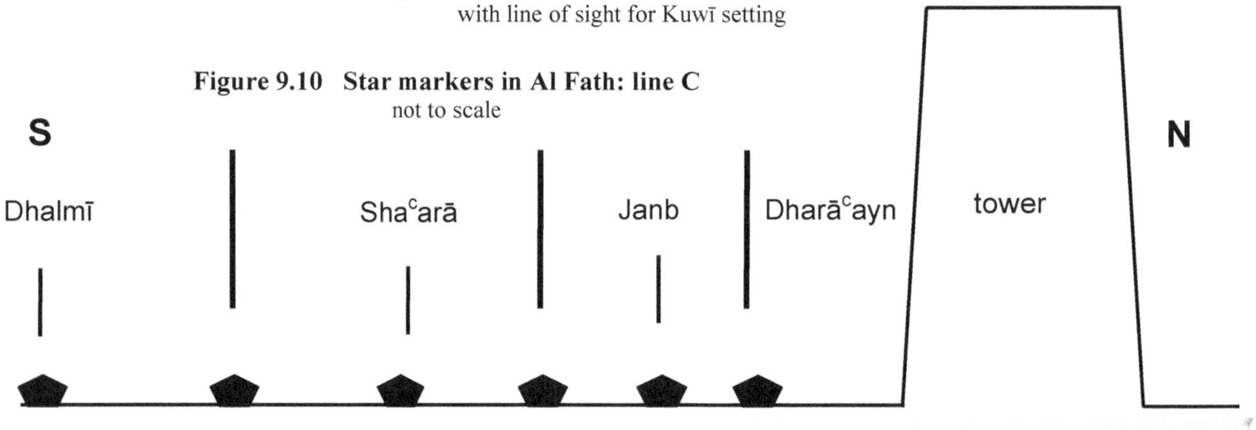

Figure 9.11 Line of sight for watching ^cAyyāq ath-Thurayyā (Capella) in Al Fath

Figure 9.12 Location D in Al Fath for watching Thurayyā (the Pleiades)
Demonstrated by Hamdun bin Humaid Al-Habsi
30 April 2006

Figure 9.13 Star markers in Al Fath: walls E and F

10 ZAHIB

10.1 INTRODUCTION

A number of tribal groups live in Zahib, which, with its *falaj*, was built in about 1910 as a satellite to Mudaybi, together with nearby Ridda and Shariq:[61] it appears that the population of Mudaybi had increased, making the expansion of agriculture and the building of new aflāj necessary. The location of the village is shown in Figure 7.3 and its layout in Figure 10.1. In the south of the area, many access shafts can be seen, which are for aflāj further downstream such as for Ridda.

The sharīᶜa (traditionally the point of first domestic use by the village) is no longer used and the washing facilities for men in the nearby mosque are said to be from well water. However, the *falaj* is still used for ablutions, bathing, and washing corpses, food, clothes and dishes (Figure 10.2). One point on the *falaj* channel between the sundial to the east and stargazing point to the west has been deepened to facilitate domestic use.

10.2 ORGANISATION OF THE *FALAJ* IN ZAHIB

The *falaj* officials in 2007 are listed in Table 10.1. The posts are decided by a group of villagers. They say that there are no disputes over the water distribution that cannot be resolved by them.

Saud bin Abdullah Al-Habsi is the *wakīl* (administrator) for both Zahib and Mudaybi, a post that has been handed down from father to son for four generations, possibly ever since the satellite villages were built. Although Khalfan bin Hamud Al-Gabri is the *ᶜarīf* (officer), he does not know about the water distribution and timing: his role is to supervise major works on the *falaj* such as repairs and extensions that involve the use of contractors, and for this he receives a percentage of the contract price. Saleem bin Rashid Al-Hashmi, a farm worker, is the most knowledgeable about the time keeping and allocation of water, though he does not have an official title. He started stargazing about seventeen years ago and water allocation in general some twelve years ago. There is no auctioneer.

The irrigation cycle is seven days, and the day starts at sunset. The *badda*s are for a full 24 hours as in Al Fath, and are known by the name of the day: they do not have special names. Each *badda* has two *farda*s (one for daytime and one for night). The term *sahm* is not used here or in Mudaybi. It is considered that time keeping with the sun and stars has been used since the beginning of the *falaj*. Saleem bin Rashid had heard of the water clock, but it has never been considered for use here. He could not remember when he got his first watch.[62]

The sundial (Figure 10.3) is used by most of the shareholders: it is well kept and has a number of personal markers, confirming this observation. It has twelve divisions each side, consistent with the division of daytime into *athar*s. The pole is steel bar, 2.4 m high.

The sundial and place for watching stars are close together, and each person is responsible for their own time keeping. When the weather is cloudy watches are used.

The time from the last star at night to sunrise, and from sunset to the first star, is two *athar*s in both cases, and like any other *athar*s may have a number of owners. A person's turn changes from day to night each irrigation cycle, falling in the same *athar* after sunset or sunrise (Figure 10.4), as in Qarya and Sudayra. Fields are watered in order of the *athar*. This means that if someone owns an area of cultivation needing two *athar*s, and they own *athar*s at different times, the watering will take place at different times. This entails some wastage of water for repeated channel wetting. If there is not enough water, half of the area is watered one week and the other half the next week.

Title	Name	Role	Started	Other work
wakīl	Saud bin Abdullah Al-Habsi	Keeping accounts; saving surplus income to pay for major works	1995	Post Master in Sinaw
ᶜarīf	Khalfan bin Hamud Al-Gabri	Supervising such works	probably before 1977	farmer

Table 10.1 Zahib *falaj* officials

[61] Sheikh Sultan bin Mohammed Al-Habsi, 20 October 2007, and Saud bin Abdullah Al-Habsi, 21 October 2007.

[62] Most information from Saleem on the *falaj* was obtained in May and November 2006.

Figure 10.1 Layout of Zahib

Figure 10.2 Food preparation in the *falaj* at Zahib
November 2006

Figure 10.3 The sundial at Zahib

The main owners of water shares are:

- Khalfan bin Hamud Al-Gabri and his brother Khuta: 32 *athar*s
- Al-Towqi family: 24 *athar*s
- Braik bin Abdullah Al-Busaidi: 20 *athar*s.

Most people own between three and six *athar*s, including some women and children. Saleem had only one *athar* in 2006, but this had increased to 2½ by October 2007. To buy one *athar* outright cost about 4000 riyals in 2006. Some people from outside the village own land and use the *falaj* water, including people living in Al Ayun near Sinaw, and Bedu from further afield - Mahut and the island of Masira, some 220 km to the south of Zahib, were mentioned by Saud al-Habsi on 26 March 2007. There are no written records of water ownership for Zahib: the *wakīl* and Saleem have a good knowledge of who owns what, and it seems that people cooperate sufficiently that written records are not needed. All sales transactions are said to be seen by and stamped by the *qāḍī* (notary/judge), but no one asked either had such a document available or was willing to let me see one. Saud's grandfather wrote a book listing the water rights of the *falaj* of Mudaybi in 1335/1917. The first page of this is given in Appendix B: it is in a standard form, stating the amount of water that each person owns. It is possible that the ownership was more complex than in Zahib, or that the system there was more often disputed to lead to the document being written.

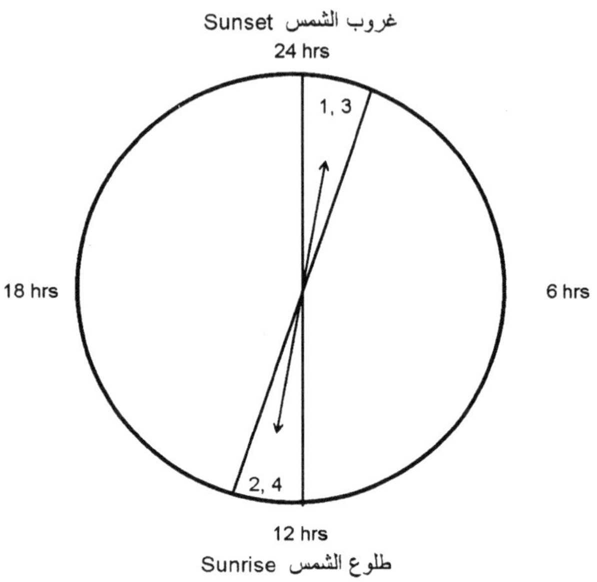

Figure 10.4 Rotation of water shares in Zahib

The *wakīl*, Saud Al-Habsi, has the most reliable information about *falaj* finances, since he keeps any surplus to pay for repairs or construction work. Although he probably keeps a written record, the information was given verbally.[63] Unlike most other villages there are no water auctions, but every year in the month of Muḥarram, the first month of the year in the Islamic calendar.[64] Saleem Al-Hashmi sells 24 *athar*s (12 hours) for 25 riyals each, raising 600 riyals for the *falaj*. He does not receive payment for this. There is an additional small income from some of the date palms along the *falaj* channel. People do not leave land or water shares to the *falaj* in their wills, but may give money to the *falaj*. A summary of the finances is given in Table 10.2.

Income		**Expenditure**	
Source	Amount riyals	Purpose	Amount riyals
annual sale	600	regular cleaning	c. 90
falaj trees	Little	10% of profits to the *wakīl*	c. 50
donations	variable	25% of contract fee to the *ᶜarīf* for supervision of works	variable
TOTAL	600	TOTAL	c. 140

Table 10.2 Summary of annual *falaj* finances in Zahib

In general, the income is more than the outgoings, the surplus being kept for major works. It may only take 80 to 100 riyals per year for regular maintenance, ie cleaning silt from the channels (and tunnel), but if there is a flood and silt enters the access shafts, much more is needed. In 2006, the main tunnel was extended by about four shafts, each costing 1,200 riyals, and in 2007 a tributary was being extended because the water was not sufficient. Funds for such major works come from wealthy people in Mahut who own date gardens here. As elsewhere, people do not pay taxes on their *falaj* property or its produce.

10.3 STAR NAMES IN ZAHIB

As with other settlements in this area there are twenty-four main stars, each covering two *athar*s. The names of the stars were given verbally on various occasions and recorded on 29 March 2008; they are listed in Table 10.3, together with their identity and English name as established during the field work for this study, with a few comments on their identification. The names are not available in written form and the Arabic is therefore given as a phonetic transcription of the spoken name. Other stars used only as indicators are:

- Shaᶜra (Sirius - α Canis Majoris), and
- *ᶜayyāq* ath-Thurayyā (Capella - α Aurigae).

Initially, it was said that Capella (α Aurigae), though used as an indicator star (see section 10.4), does not have a name, but later that it is *ᶜayyāq* ath-Thurayyā, implying that the term *ᶜayyāq* is a description rather than a name. The locations of the stars in the sky are shown in Figure 10.5.

Stars for which the identification is uncertain are shown in the table in italics. There are more uncertainties here than in the other study villages, partly because Saleem was not comfortable with looking at the computer, and that means of checking identifications was therefore not available, but also due to bad luck with the weather, with several hazy nights limiting the number of stars that could be seen. In March and April 2006, Thaqīlā, pointed out as rising a little north of Dharāᶜayn, or the brighter star used as an alternative should have been visible, but light cloud cover meant that it was not, and on visits in March and April 2008 the sky was also cloudy. Thaqīlā rises east-north-east, not far enough north to be one of the stars by this name formerly used in Sudayra. The stargazers also could not be sure about the alternative star, giving failing eye sight and light pollution as the reason.[65] However, the heliacal rising of Thaqīla is associated with the beginning of the season for agriculture (Saleem Al-Hashmi, 18 October 2007).

[63] mostly on 26 March 2007.

[64] A list of Islamic calendar months is given in Appendix C.

[65] Saleem bin Rashid Al-Hashmi and Braik bin Abdullah Al-Busaidi at different times.

Star name as spoken by Saleem Al-Hashmi		Number of stars	Magnitude	English name*	Identity*	Comment
Arabic	**English**					
الصارة الاولية	Ṣāra (*al-awliyya*)	1	2.43	Scheat	β Pegasi	
الصارة الثانية	Ṣāra (*ath-thāniyya*)	1	2.81	Algenib	γ Pegasi	definitely not Alpheratz
السعد	Saᶜd	1	2.06	Mirach	β Andromedae	
الكوكبين	Kawkabayn	2	2.62 2.00	Hamal + Sheratan	α + β Arietis	
الفتح	Fatḥ	1	3.59		41 Arietis	not very bright, rises close to above
الثريا	Thurayyā		≥ 2.84	Pleiades	Messier 45	
الدبران	Dubrān	1	0.84	Aldebaran	α Tauri	
الشابك	Shābik	3	3.37 – 4.37	Meissa (λ)	$\lambda + \varphi^1 + \varphi^2$ Orionis	
البلمي	Balmī	1	1.90	Alhena	γ Geminorum	
الشعرة	Shaᶜra	1	0.37	Procyon	α Canis Minoris	
الجنب	Janb	c. 2	3.37		ε Hydrae +1	use most visible
الذراعين	Dharāᶜayn	2	1.34 2.00	Regulus + Algeiba	α + γ Leonis	
البطين	Boṭayn	1	3.31	Chertan, Chort	θ Leonis	
الثقيلة	Thaqīla	1	≥ 4.34		*in Coma Berenices*	
الميثاب	Mīyathāb	1 or 2	2.84	Vindemiatrix	ε Virginis + 1	use the most visible
الذكرين	Dhakarayn	1	-0.07	Arcturus	α Boötis	
الغفر	Ghafar/Aghfar	1	2.21	Alphecca, Gemma	α Coronae Borealis	bright
الزبان	Zibān	1	*2.75*	*Kornephoros (β)*	*β* or *ζ Herculis*	bright, rises below previous
الاكليل	Iklīl	1	*3.09*	*Sarin*	*δ Herculis*	rises a little N of Arcturus
لكوي	Kuwī/Kuwiyya	1	0.00	Vega	α Lyrae	
المنصف	Munṣif		3.03	Albireo	β Cygni	
النسر	Nisar	1	0.75	Altair	α Aquilae	
لغراب	Ghurāb	4	3.62 – 4.40	$\alpha, \beta, \delta + \lambda^2$ Delphini	Sualocin (α), Rotanev (β)	
اللدم	Ladam		2.37	Enif	ε Pegasi	
* *italics*: identification uncertain.						

Table 10.3 *Falaj* stars identified in Zahib

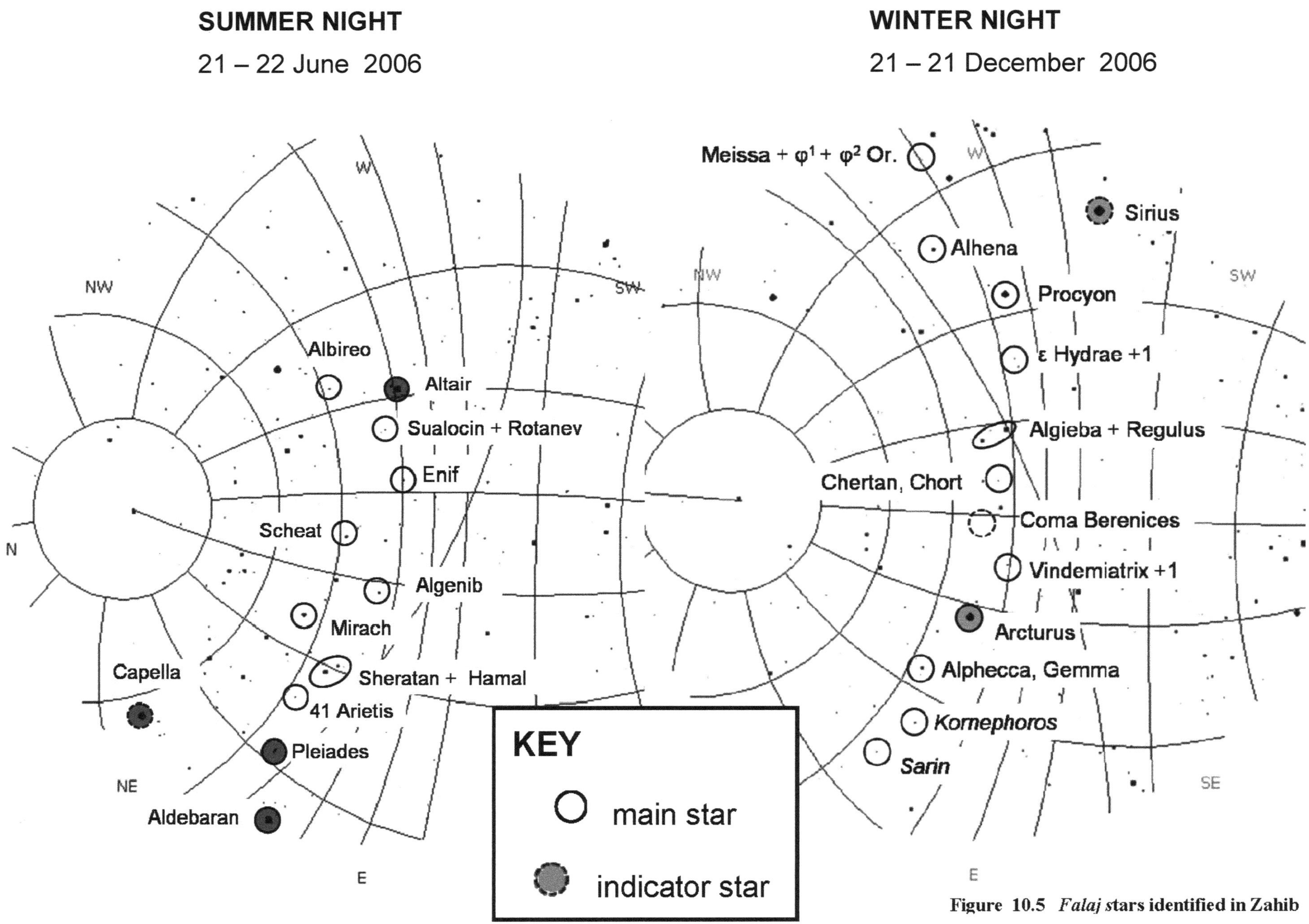

Figure 10.5 *Falaj* stars identified in Zahib

10.4 METHOD OF WATCHING STARS IN ZAHIB

The place for watching the stars is near the sundial. According to Saleem bin Rashid, everyone uses the stars but checks with him when they are unsure, and when it is cloudy, wristwatches are used. As in Al-Fath, an indicator star is selected to mark the rise of one or more of the main stars listed above, so the farmers do not need to be able to recognise the main star. This might explain why it was not possible to firmly identify all of the main stars.

The walls on both sides of an alley oriented east-west are used (Figure 10.6). The walls on the south side belong to a ruined house said to be about 140 years old; the wall on the north is for a garden adjacent to an inhabited house. There are three zones for watching from:

- A the wall of the house on the south side of the alley and east of the pole, looking west (and slightly north) – the wall of the house is the marker;
- B south of the alley to the west of the pole – short wall, looking east – the horizontal pole is the marker;
- C north of the alley looking south east – the tower of the house is the marker.

It appears that nearly all of the marks in these zones are used. However, the markers and post on the new part of the north wall, in front of area C on Figure 10.6 are not used: Braik bin Abdullah Al-Busaidi (10 May 2006) thinks it possible that children had been trying to copy the system. Each person watches for their own star, as illustrated in Figure 10.7, and several expatriate farm workers have learned to use the system.

The locations of the indicator stars in the sky are shown above in Figure 10.5 and their names, the main star they are used for, and the observation points are listed in Table 10.4. As in Al Fath, there are two *athar*s and two marks to each star: the beginning of the first and second *athar*s, while the end of the second *athar* is the first mark for the next star. Hence the numbers 1 and 2 after a few stars in Table 10.4 refer to the first and second *athar*s respectively.

Figures 10.8 to 10.14 below show the observation points, mostly vertical scratches or small hollows in the wall, and the main stars for which they are used in the order than an observer would follow. We start at wall A (Figure 10.8) at the beginning of the time of Kawkabayn (Hamal and Sheratan) and move eastwards, watching the indicator stars setting over the corner of the house to which the horizontal pole is attached; but it is the wall that is used to mark the setting, not the pole. The last observation point on this round is for the start of the second *athar* of Balmī (Alhena, α Geminorum).

Figure 10.6 Alley used for stargazing in Zahib

Figure 10.7 Saleem bin Rashid Al-Hashmi and the author waiting for Shaᶜra (Sirius) to rise

Main star	Star watched	Point watched from	Marker for rise/set
Kuwī	Dhakarayn	B	rise above pole
Munṣif			
Nisar			
Ghurāb 1			
Ghurāb 2	Dhakarayn	A	set behind wall of house
Ladam			
Ṣāra (*al-awliyya*)			
Ṣāra (*ath-thāniyya*)			
Saᶜd			
Kawkabayn 1	Kuwī	A (east wall)	set behind wall of house
Kawkabayn 2		A	
Fatḥ			
Thurayyā			
Dubrān			
Shābik	Nisar		
Balmī			
Shaᶜra (Procyon)	Dubrān	B	rise above pole
Janb			
Dharāᶜayn	Shaᶜra (Sirius)	C	rise above tower
Boṭayn 1			
Boṭayn 2	Shaᶜra (Procyon)	B	rise above pole
Thaqīlā			
Mīyathāb			
Dhakarayn	*ᶜayyāq* ath-Thurayyā	A	set behind wall of house
Ghafur/Aghfar			
Zibān	Dhakarayn	B	rise above pole
Iklīl/Liklīl			
Where the main star is given twice, followed by 1 or 2, 1 refers to the first and 2 to the second *athar*.			

Table 10.4 Indicator stars used in Zahib

The next marker is at the west end of wall B, marking the end of Balmī and the beginning of Shaᶜra (Procyon; Figure 10.9), with the indicator star Dubrān (Aldebaran) watched rising over the pole. After Janb we cross the alley to area C for the first marker for Dharāᶜayn. There are only three observation points on this wall and the indicator star Sirius is watched rising over the top of the tower of the house (Figure 10.10).[66]

For the beginning of the 2nd *athar* of Boṭayn, we move back to wall B, again starting in the west and working eastwards towards the pole, to the marker for the beginning of the 2nd *athar* of Mīyathāb (Figure 10.11) and then to A (Figure 10.12), B (Figure 10.13) and A (Figure 10.14) again. This may seem complicated, but on any one night, an observer may only need one of the points on one wall.

When asked how he thought the stars were watched before the buildings now used were present, Saleem said, as is indeed the case, that any building can be used for this purpose. He thought that originally, people would have simply watched the main stars rising above the horizon but developed the current method when the horizon became less clear. However, the question was asked before realising that Zahib is a relatively recent settlement. It is most likely that Zahib copied the system from Mudaybi, where people stopped using stars when the mosque used for watching them was repaired in about 1990, since when the few elders who were knowledgeable about the stars have died. The former use of a building indicates that the system of stargazing was similar to that in Zahib and Al Fath.

[66] This wall was demolished in 2008 and wristwatches are now used for this section.

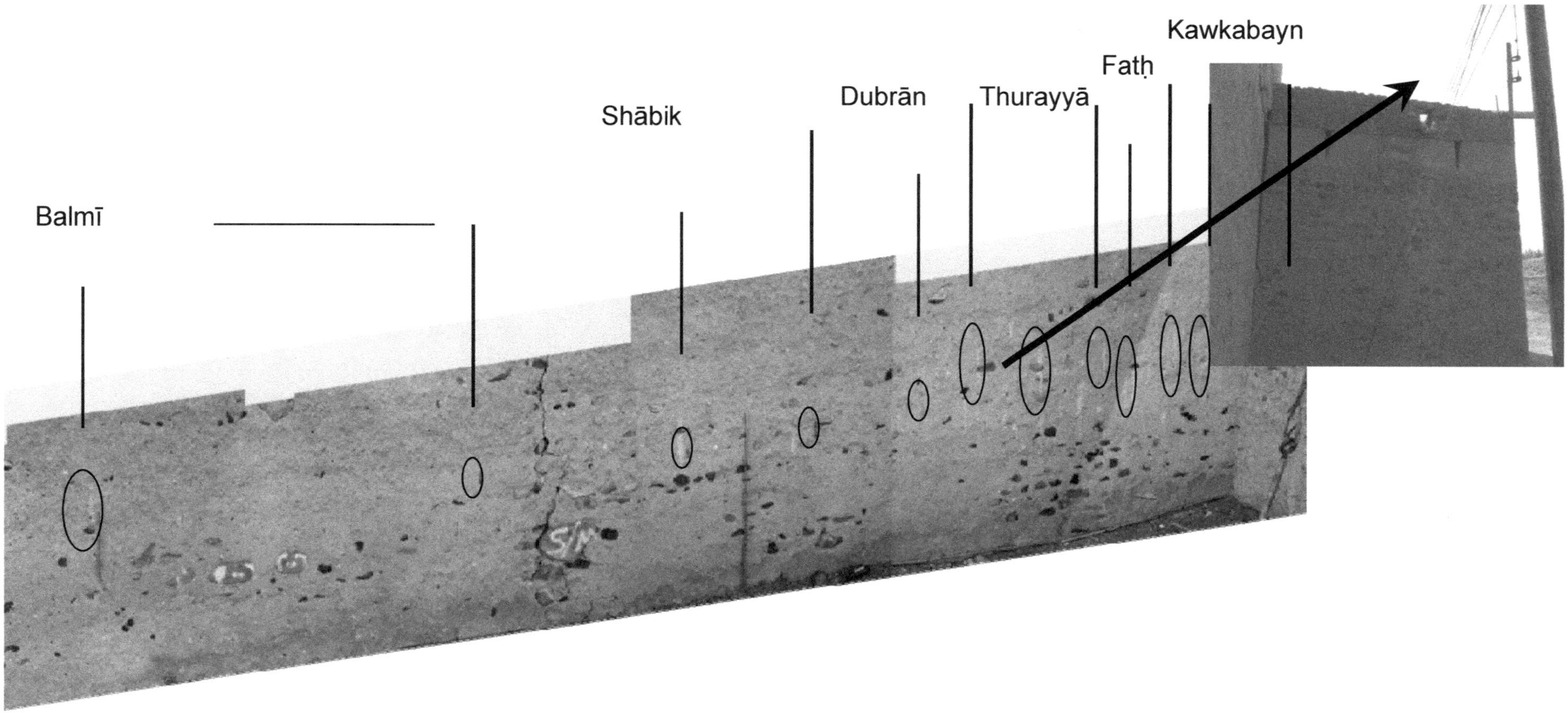

Figure 10.8 Star markers in Zahib: wall A round 1

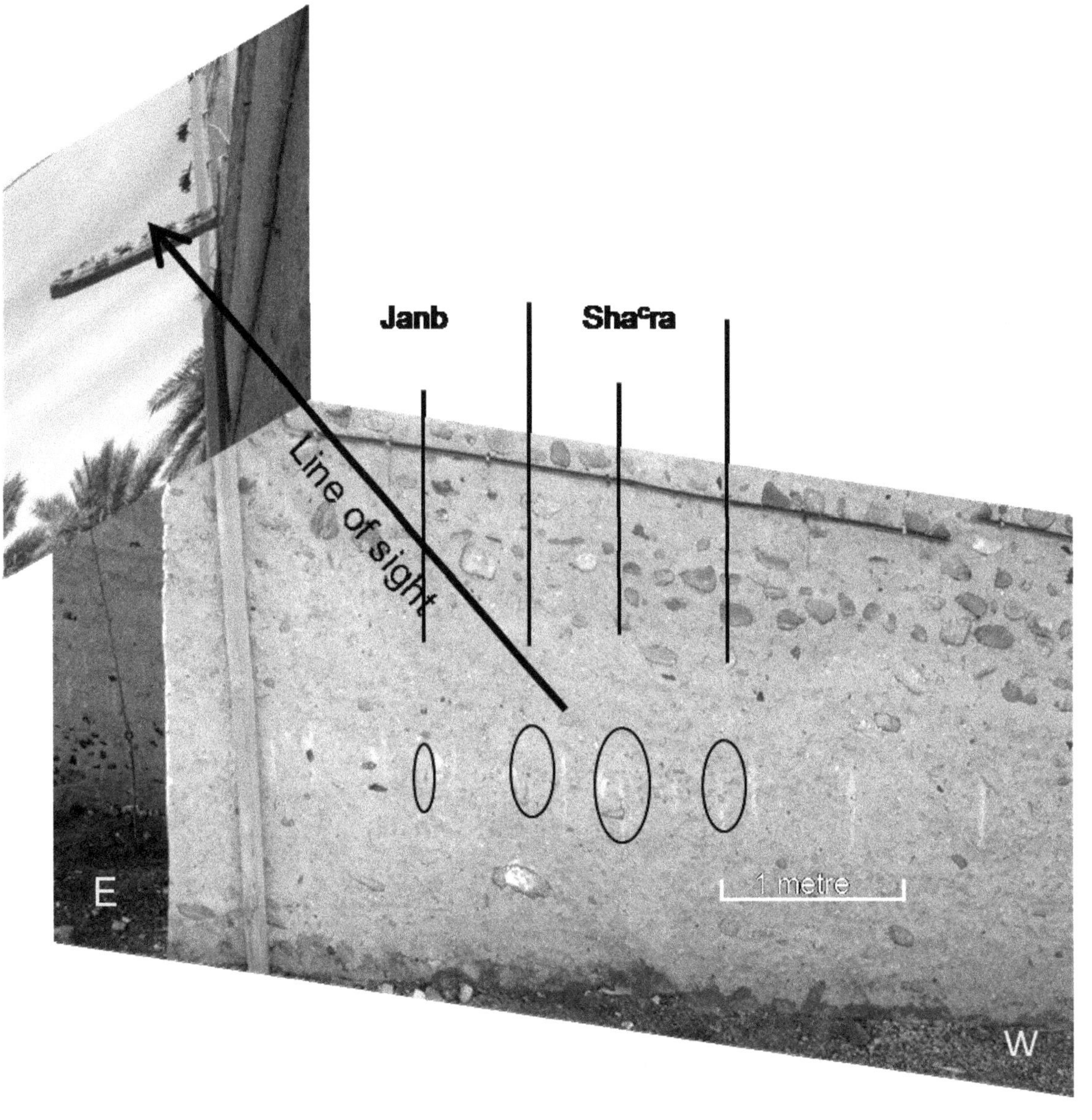

Figure 10.9 Star markers in Zahib: wall B round 1

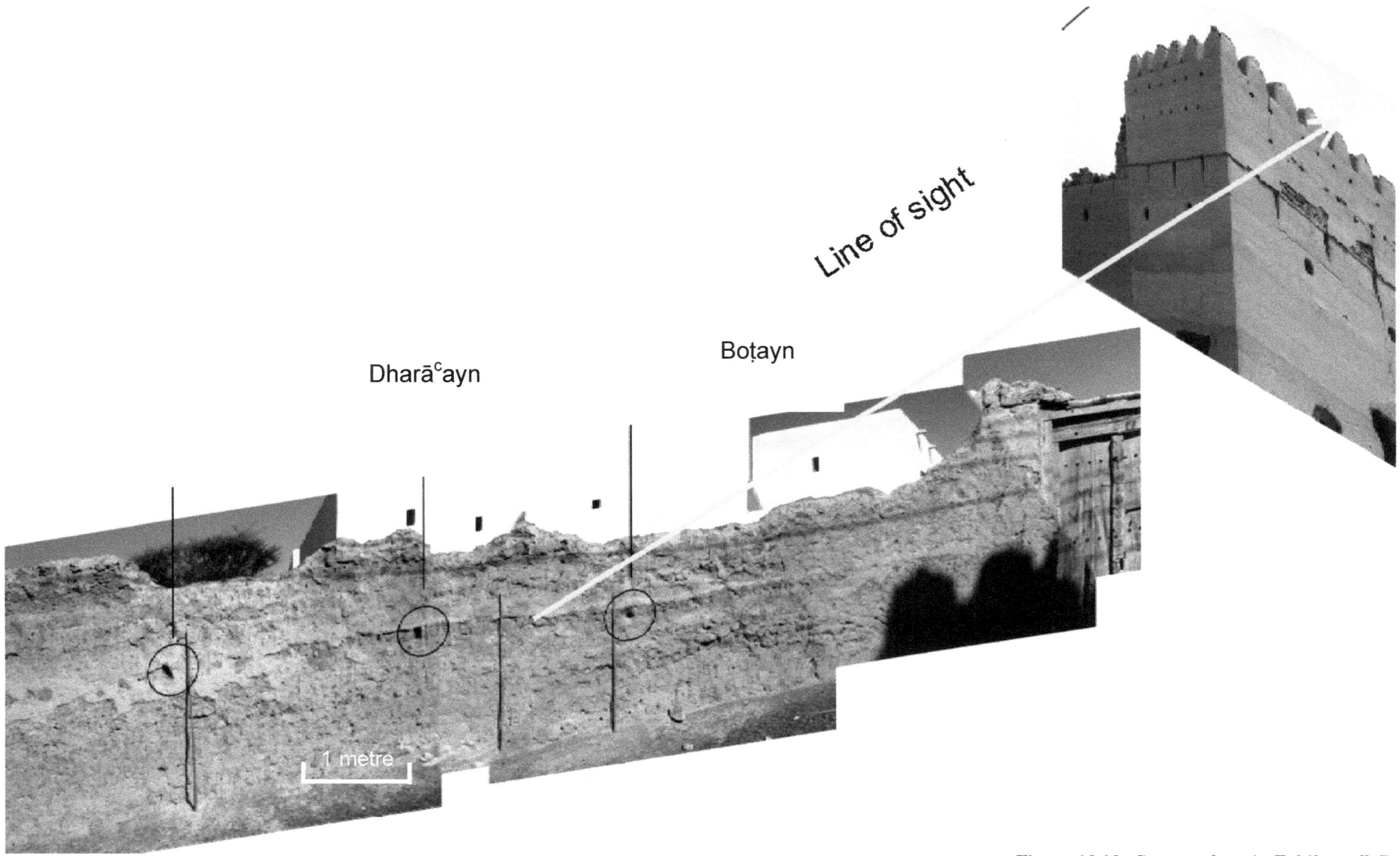

Figure 10.10 Star markers in Zahib: wall C

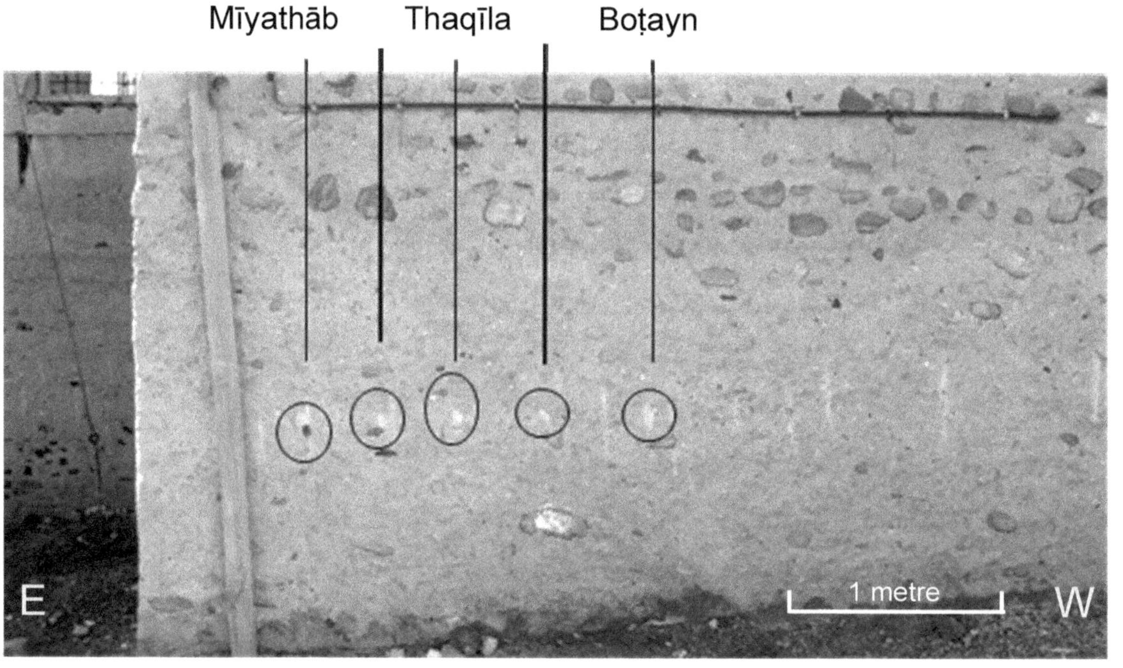

Figure 10.11 Star markers in Zahib: wall B round 2

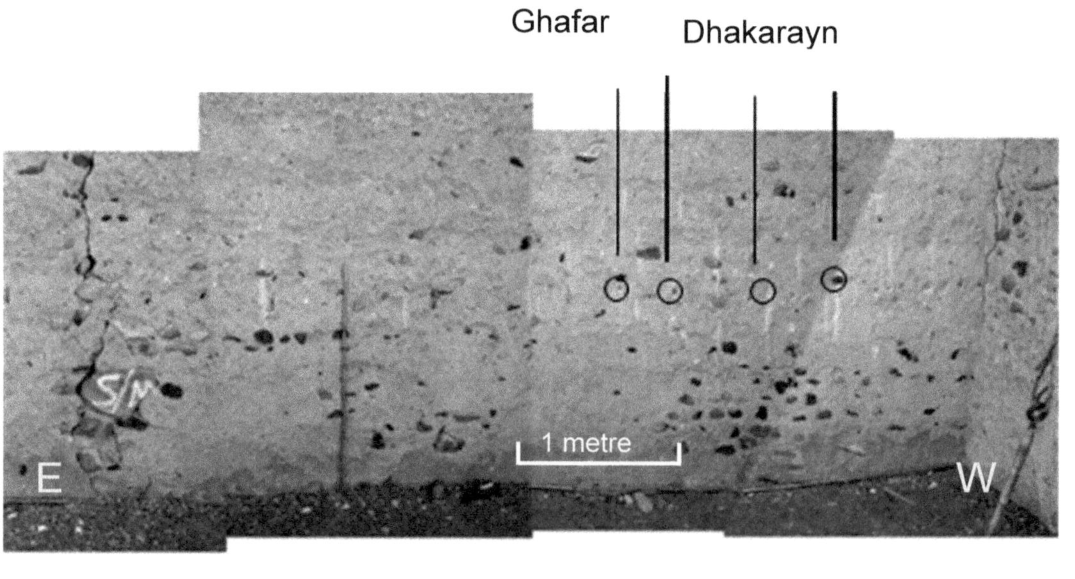

Figure 10.12 Star markers in Zahib: wall A round 2

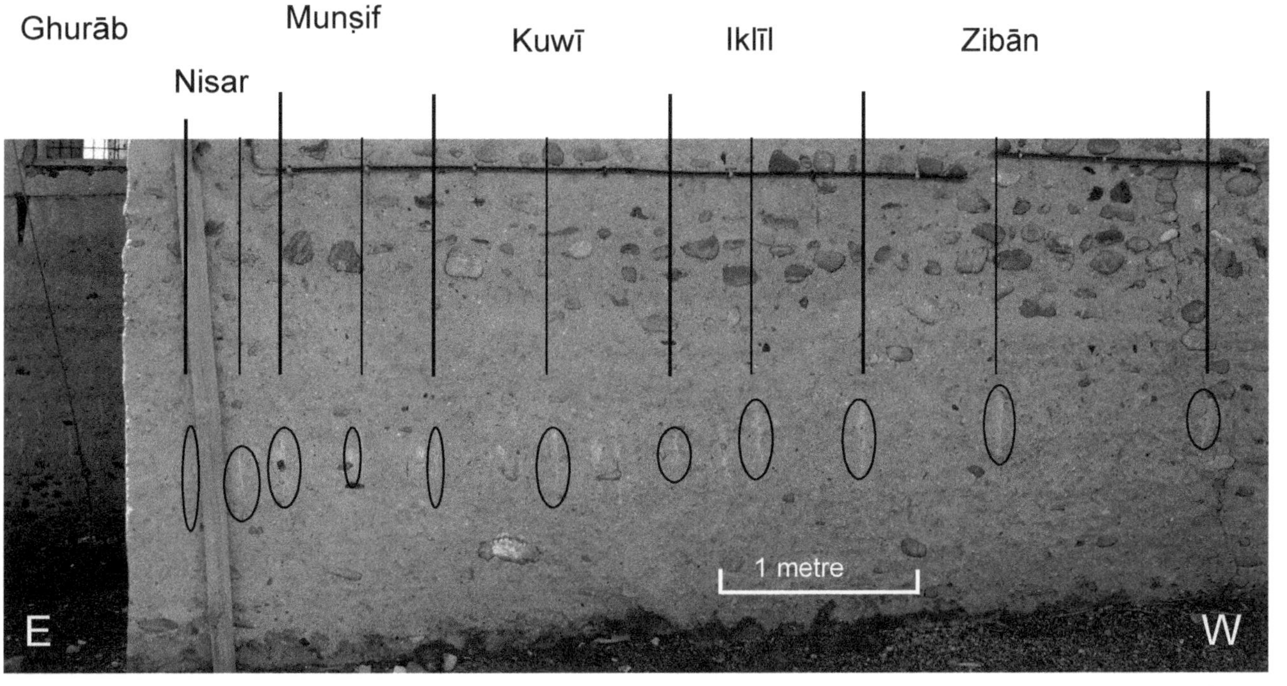

Figure 10.13 Star markers in Zahib: wall B round 3

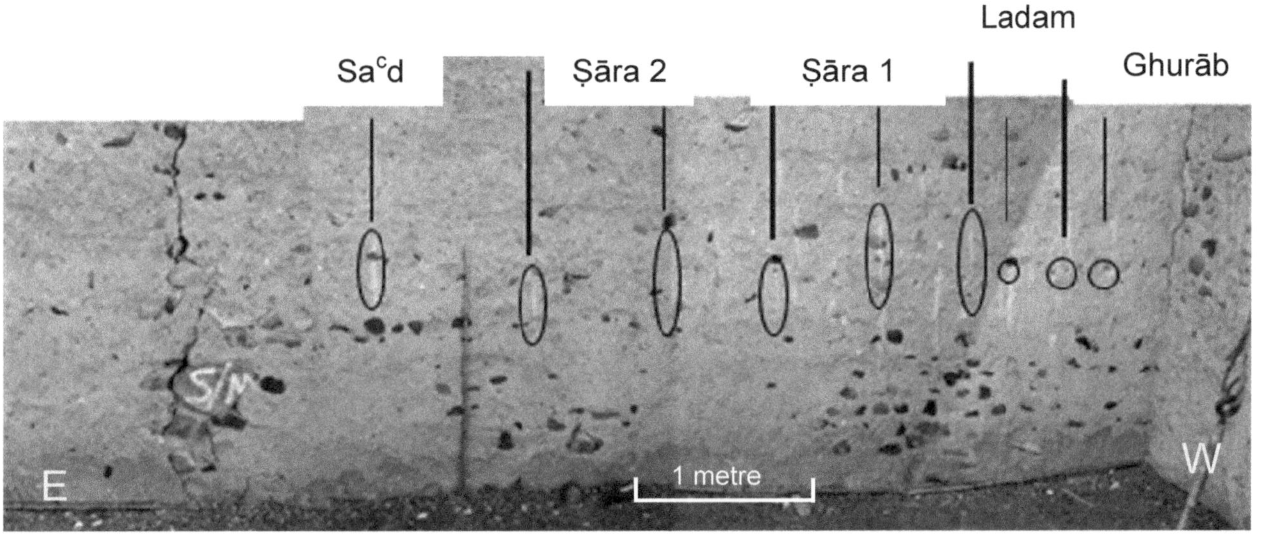

Figure 10.14 Star markers in Zahib: wall A round 3

11 DISCUSSION

This chapter draws together the information collected in the study villages and relates it to information collected elsewhere. It follows approximately the same order as the sections of the study areas/villages, starting with *falaj* organisation followed by time keeping and stars. The selection of stars is given particular attention as well as their names and variations from place to place.

11.1 *FALAJ* ORGANISATION

The study villages were selected because the stars are still used there and they are therefore not representative of modern communities as a whole. However, information was also collected from a few villages no longer using stars, and provided that the bias of the data is kept in mind, some valid comments can be made of more general applicability.

It is clear from the information gathered in the field that the organisation and management of the *falaj* varies considerably from one place to another. The responsibilities of the officials fulfil the traditional demands, but may be shared between the *ᶜarīf* (officer) and *wakīl* (administrator), indeed there may be no *ᶜarīf*, as in Misfa, or two of each as in Barzaman. The use of two *wakīl*s in Qarya and Barzaman appears to be an adjustment to modern life, in which several officers have paid employment unrelated to *falaj* activities and are therefore not always available. The case of Sudayra, where the system seems to have broken down is the exception, at least for the present. Auctions of water for *falaj* maintenance are no longer *de rigueur*, but water is still rented or sold for this purpose. However, the difference to former practice is that the funds raised are not sufficient to fully maintain the system: without private donors and government support, many more *falaj* systems would have stopped working. The alternative, of private wells, does not need or contribute to the sense of community and society that is integral to the traditional system.

The responsibility for timing the *falaj* water depends on cooperation among the farmers, and on a daily basis each is responsible to see that they take their water at the allotted time. The time divisions where the sundial and stars are still used are variable: given that the system is based on a basic division between day and night, a certain amount of variation is to be expected. The length of each at this latitude ranges from approximately 11 to 13 hours, so the *athar*, at 24 to either day or night should range from 27 to 33 minutes. The variation seen, at least at night in places using the horizon, is much greater than this but will even out over the year. Equalisation is also achieved by alternating the shares between different parts of the day and night, as in Barzaman and Sudayra. In other places, such as Qarya, a higher value is placed on longer shares.

11.2 DOCUMENTATION

Information on star names, their timings and the method of watching them is passed down orally from generation to generation, so there is very little written about them. Although water rights and transactions are recorded in *falaj* books in some settlements, none of those seen contain any information about the method of timing.

The manuscript from Al Hamra discussed in section 6.4.3 is an exception, with a fairly detailed description of each star and the time it represents. Although the method of watching is not described, the descriptions clearly apply to watching the horizon. However, if one such manuscript exists, albeit only in later copies, there may be others. Towards the end of the study, I found that Sheikh Mohanna bin Khalfan Al-Kharusi from Awabi, near Rustaq, has a collection of manuscripts, including some on stars. The librarian of the Sayyid Mohammed bin Ahmed Al-Busaidi library near Seeb kindly gave me one of Sheikh Mohanna's books: *Al-nujūm al-zāhira fī l-aflāk al-rā'ida* (Bright Stars in the Celestial Sphere) (al-Kharūṣī, 1465/2004). This has section of sayings related to weather and agriculture for each station of the moon, but his sources are not cited. Sheikh Mohanna met me on 28 November 2008 and confirmed that 1) he does not have any original manuscripts on stars from Oman, and 2) that the information on stars in his books is from earlier works such as Al-Ṣūfī's (section 5.3).

Also there are the star lists for some of the study villages, already described, and lists provided by other authors for the villages they were working in. The latter are referred to below in the discussion on the selection of stars in the study area.

Apart from the above, I saw and photographed two star charts, one from Mudayrib in the Sharqiyah Region, between Mudaybi and Mintirib on the road to Sur, and the other from Mudaybi. When investigating the former use of stars in Mintirib on 9 January 2006, the Sheikh and *wakīl* (administrator) of the *falaj*, Sultan bin Rashid Al-Hajri took me to meet Hamad bin Saeed al-Harthi (d. 2006), who had painted the chart shown in Figure 11.1 in c. 1986. He had done this for his own interest, not for the purpose of *falaj* management The chart is entitled "*Dawrān al-aflāj*" (Cycle of the *Aflāj*) and presents the names of the *falaj* stars formerly used in Mudayrib, showing which season each was used in and the time in minutes (in the stars around the circumference) between each. The time between any two stars ranges from 20 to 80 minutes, similar to the variation in some of the villages in the study areas. The star names are similar to those in the Mudaybi area but there are some differences, described clockwise from the top:

- Janb (3rd down in the bottom left quadrant) is written al-Yanb, as the name would be spoken in many places;

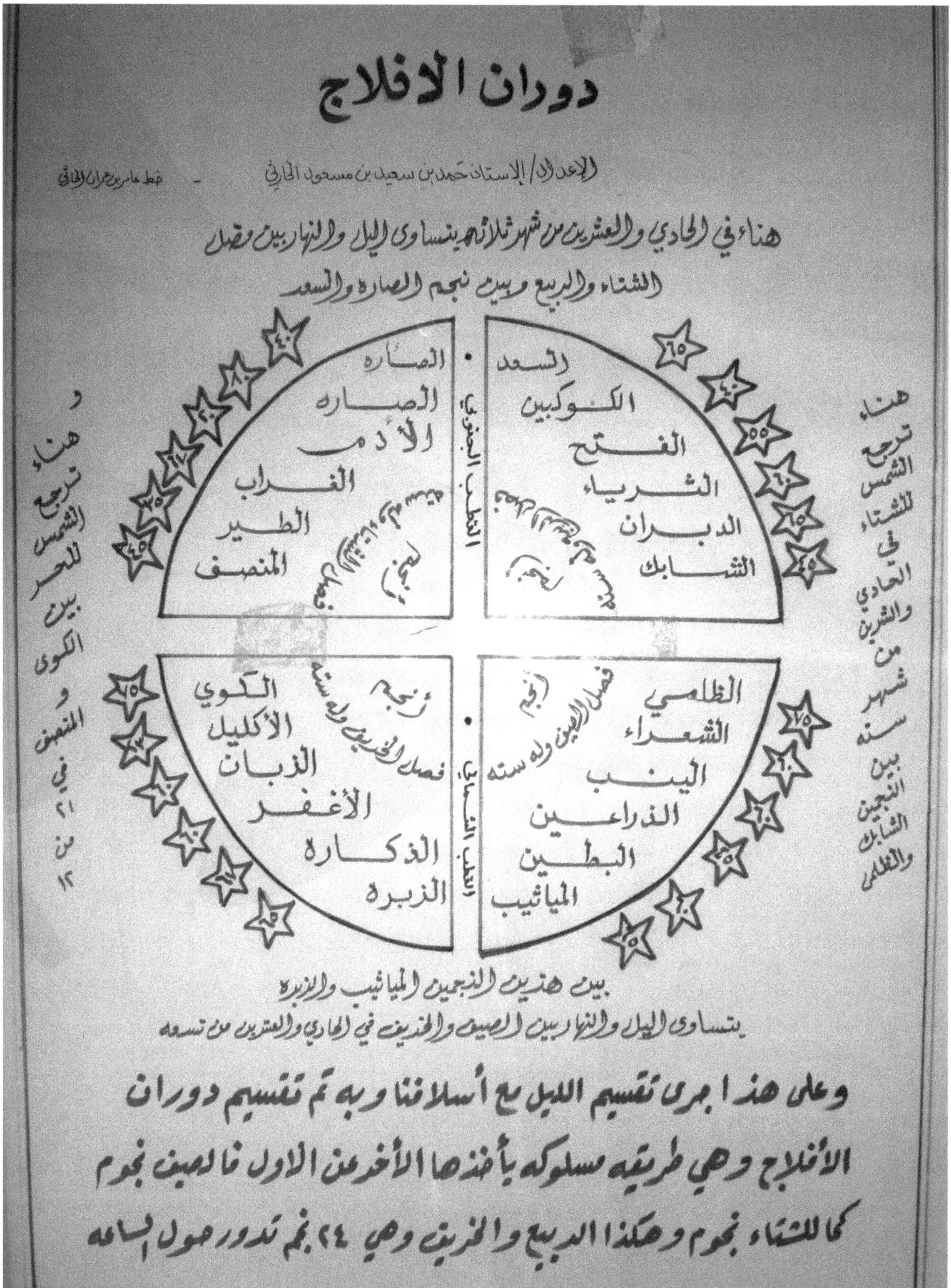

Figure 11.1 Star chart from Mudayrib
from Hamad bin Saeed Al-Harthi, 9 January 2006

- Thaqīla is not present between Buṭayn and Mīyāthīb (Mīyathāb/Mawāthīb elsewhere);
- another star not used in the Mudaybi area, Zubra, is present (lowest in bottom right quadrant of the chart) between Mīyāthīb and al-Dhakāra;
- the star(s) elsewhere known as Dhakarayn is here Dhakāra;
- Ṭayr (the bird, more formally Ṭā'ir) takes the place of Mūfī (usually identified as Altair, α Aquilae) or Nisr.

The timing before and after Ẓalmī, the first star in the bottom right quadrant, assisted in the initial identification of this star in the study villages in the Mudaybi area as Alhena, γ Geminorum, the same as Jawza' in Qarya, later confirmed by field observations.

Saud Al-Habsi's (the current *wakīl* – administrator for the *aflāj* in both Mudaybi and Zahib) late uncle, Hamud bin Ahmed bin Rashid Al-Habsi, drew up a star chart in Mudaybi, probably in the 1960s (Figure 11.2).[67] This is entitled, in the inner circle, *dawr al-falak* i.e. the astronomical cycle. From the centre, the second circle gives the four seasons; the third gives the twelve signs of the zodiac; the fourth is the stations of the moon, and the outer gives the *falaj* stars. This appears to have been an attempt to relate the *falaj* stars to stations of the moon (see section 11.5 below), but the list of *falaj* stars may be incomplete in that Ghafr is repeated and Zubān is missing. Again it appears that this was drawn for the interest of the artist, not for use by stargazers or the *falaj* administration.

None of these documents except perhaps the Al Hamra manuscript attempts to give sufficient detail for the system of stargazing to be learned. It must have been written at about the same time that Misfa started using stars, but my sources in Misfa had never heard of it or seen a copy, so it is unlikely to have been written for their ancestors.

11.3 IDENTIFICATION OF STARS

Since the knowledge of *falaj* stars is an oral tradition, in all of the villages the stars are identified by observation: by their number, relative brightness and their location of rising or setting and in relation to each other. The star colour appears not to be taken into consideration, not being mentioned in speech except when asked if a star is red or white, when the answer indicated an awareness of the colour. Nor is the colour mentioned in the manuscript from Al Hamra. However, in Stall in Wadl Beni Kharus near Rustaq, on the north side of the mountains from Al Hamra (visited from 2 to 3 April 2008), both Procyon (α Canis Minoris) and Sirius (α Canis Majoris) are used as main *falaj* stars and are distinguished by name as

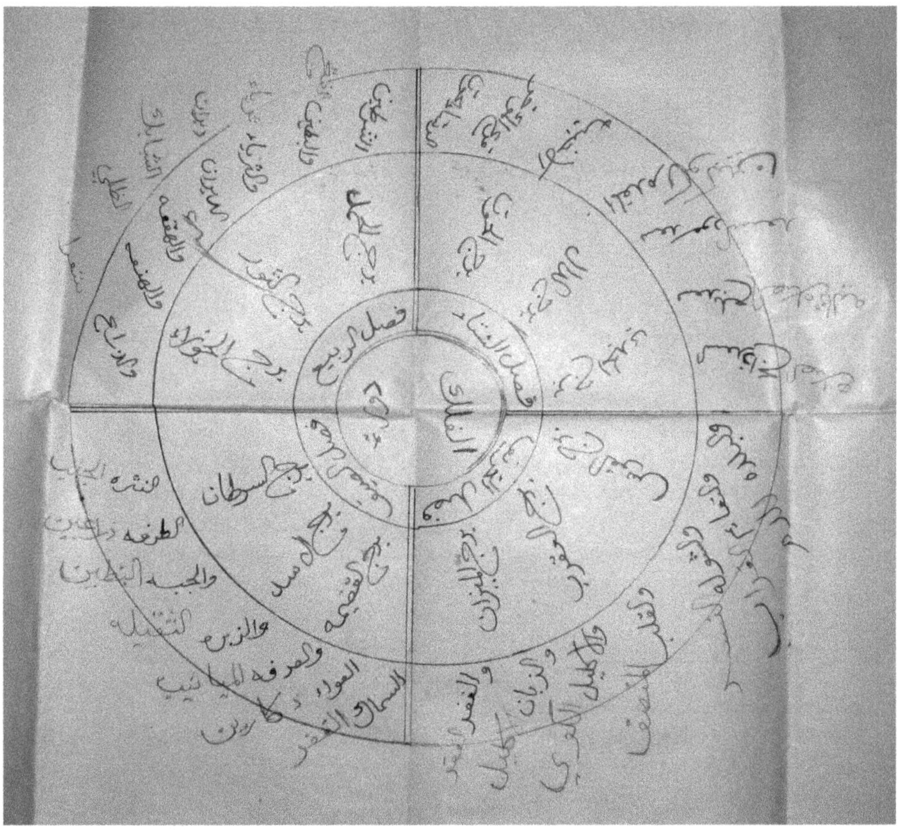

Figure 11.2 Star chart from Mudaybi

[67] Saud bin Abdullah Al-Habsi, 21 October 2007.

Sha^cra al-Ḥamra (red) and Sha^cra al-Bayḍa (white) respectively.[68]

The identification in relation to other stars is important: in Barzaman, for example, many of the stars have an ^{c}*ayyuq*, mostly rising up to c. 25 minutes after the main star. Only for Kuwī (Vega, α Lyrae) does the ^{c}*ayyuq* (Antares, α Scorpii) rise at the same time. This appellation seems to mean a star accompanying the main star and used as an indication of when the rise of the main star has, with the one exception, already happened. The use of pairs of stars often some distance apart, as in Qarya, serves a similar purpose should part of the sky be cloudy.

11.4 METHODS OF WATCHING STARS

The study has identified a range of methods of stargazing:

- looking in the sky from anywhere and knowing the time, as currently in Sudayra;
- watching the natural horizon: Qarya area and formerly Sinaw;
- watching a local (manmade) horizon: Barzaman and Sudayra;
- using indicator stars to represent the timing of main stars in Al Fath and Zahib.

It appears that the first method is only used in the study area where the formal system of stargazing is used rarely and only by a few people. However it was also used in settlements where there were relatively few divisions of water, as in Al Sharja, a small village inland from Sohar now practically abandoned.[69]

Of the study villages, only Qarya uses the natural horizon. However it is clear that this method was once much more widely used, as in the rest of the Al Hamra area, and formerly in Sinaw, between Mudaybi and Barzaman. It is still used in Stall and probably other villages in Wadi Beni Kharus, and in Halam, at the top of Wadi Beni Jabar, near Sur.[70] Without information from places other than the study villages it would have been reasonable to conclude that this method was used in a limited area, but this is clearly not the case. It helps to have a horizon relatively close to the observer such as that at Qarya (Figure 6.8), since the rise or setting of stars can readily be observed. The choice of watching stars rise or set is a function of which horizon is clearest: in Halam in Wadi Beni Jabar, for example, they are watched setting, as the horizon to the east is obscured by vegetation.

The use of a local manmade horizon, as in Barzaman and Sudayra, can be seen as a development of the first method for flat terrain where the natural horizon is distant or obscured. However, in these two villages there is a further modification: the observer moves from one marker to another giving more control over the divisions of time.

The complex methods used in Al Fath and Zahib appear to be a further development. The use of markers, mainly on walls, and of indicator stars means that much shorter intervals of time can be achieved with greater accuracy, and only a few stars need to be recognised. One can envisage this method developing in an urbanised area on a *falaj* system with many small divisions of water ownership. For this method of stargazing, the main *falaj* star does not need to be seen, yet my sources in those two villages, Saleem Al-Hashmi and Sultan Al-Shidhani clearly know and recognise all of them, a fact which supports the argument that originally the stars would have been watched rising above the horizon in these villages.

While assisting me in the Mudaybi area, Ibrahim Al-Namani from the Ministry of Water Resources (MWR) suggested that the selection of the method of stargazing could be a function of the complexity of the tribal make-up of the village, with a more complex system being used in places with many different tribal/family groups, such as Al Fath, and simpler systems in places with only one tribal/family group such as Barzaman.[71] Certainly, this hypothesis reflected the facts available to us at the time. However, on further investigation it was found that settlements such as Al Hamra and Qala have several tribal/family groups, yet followed the simplest system of watching stars rise above the horizon, and the complexity of tribal make up does not appear to be a major factor in the selection of method.

11.5 SELECTION OF STARS

The factors expected to be taken into consideration when selecting stars to use for *falaj* purposes are:

- they should be visible and rise and set: circumpolar stars do not rise or set and so cannot be used for time reckoning in the methods described above;
- brightness;
- approximately equal time intervals (depending on the method of watching);
- stars already known and possibly used for other purposes.

The number of main stars used varies: 21 stars are used in the Qarya area, most with two *athar*s, and a few with three *athar*s; 24 stars are used in the Mudaybi area, each covering two *athar*s. Whatever the number or spacing of stars, together they cover a full year, but the spacings are not equal and the *athar* length varies more than the variation in length of night. Table 11.1 presents a list of the stars used in the two study areas, simplified from the information given in the previous chapters by using the same spelling. The stars in bold in the Qarya area are those for 3 *athar*s: the rest are for 2 *athar*s.

68 Khalifa bin Khalfan Al-Hattali in Stall, 2 – 3 April 2008.

69 Sultan bin Saif Al-Muqbali in Muscat, 3 November 2007.

70 Visited from 12 to 13 April 2008.

71 During a visit to Zahib, 3 May 2006.

Falaj star				
Arabic	**Qarya area**	**Mudaybi area**	**Identity**	**Magnitude**
الكوكبين	**Kawkabayn**	Kawkabayn	(α +) β Arietis	2.00
الفتح		Fatḥ		
الثريا	Thurayyā	Thurayyā	Pleiades M 45	>=2.84
الدبران/بو قابل/المقبل	Dubrān	Dubrān/Bū Qābil/Muqbil	α Tauri	0.84
الشابك	Shābik	Shābik		
الظلمي/الجوزاء	Jawzā'	Ẓalmī	γ Geminorum	1.90
الشعرى	Sha^{c}rā	Sha^{c}rā	α Canis Minoris	0.37
الجنب	Janb	Janb	δ, ε + ρ Hydrae	3.37
الذراعين	Dharā^{c}ayn	Dharā^{c}ayn	α + γ Leonis	1.34
البطين	Buṭayn	Buṭayn	δ Leonis	2.50
السفاع/الثقيلة	Sufā^c	Thaqīla		
المواثيب	Mawāthīb	Mawāthīb	ε Virginis	2.84
الذكرين	**Dhakarayn**	Dhakarayn	α Boötis	-0.07
الغفر	Ghafar	Ghafar	α Coronae Borealis	2.21
الزبان		Zubān		
العقرب/الاكليل	**^{c}Aqrab**	Iklīl		
كوي	**Kuwī**	Kuwī	α Lyrae	0.00
المنصف		Munṣif		
الموفي/النسر	Mūfī	Nisr or Mūfī		
الغراب	Ghurāb	Ghurāb	α, β, δ + λ^2 Delphini	3.62
الادم	Adam	Adam	ε Pegasi	2.37
الصارة	Ṣāra	Ṣāra		
الصارة	Ṣāra	Ṣāra		
السعد	Sa^{c}d	Sa^{c}d	β Andromedae	2.06
all stars are for 2 *athar*s except those in bold, which are for 3 *athar*s				

Table 11.1 Comparison of *falaj* stars in the study areas

The identification and brightness (of the brightest star) are given only for those stars that are practically the same in each place. The locations of all of the main stars, and the indicator stars watched in Zahib and Al Fath, are shown in Figure 11.3. The furthest north are the stars in Cassiopeiae: Caph (β Cassiopeiae at 59°12'N), the first Ṣāra in Qarya, and Schedar (α Cassiopeiae at 56°32'N) with γ (at 60°46'N) +/ε Cassiopeiae as the first Ṣāra in Sudayra. Stars further north (Polaris and other stars in Ursa Minor, Ursa Major, Draco, Cepheus and Camelopardalis) are not used because they do not rise or set. The southernmost main *falaj* star identified so far comprises β and δ Scorpii, sometimes with π Scorpii (the furthest south at 26°8' S), called Iklīl in the Mudaybi area and ^{c}Aqrab in Misfa and Al Hamra. However, some divider stars are further south, such as Canopus (Suhayl, α Carinae) at 52° 42' S, which was used in Misfa.

The stars that are more or less the same in each place, listed in Table 11.1, are generally bright, with a magnitude less than 3, except for two readily recognisable groups: Janb, the head of Hydra, and Ghurāb – much of the constellation of Delphinus. If stars in other places are taken into account, the number that are the same in every place reduces. In Sual, for example (Al-Siyabi, 2000: 24-25), there are four different stars between Sha^{c}arā and Mawāthīb: ^{c}Abdayn (the two slaves, unidentified, but Dhirā^{c}ayn would be a good contender being two stars). Three stars are named the fourth, sixth and seventh, which, on checking with the author, were confirmed as stars from the Plough. In Sinaw, it was said that stars from the Plough were also used between Sha^{c}arā and Mawāthīb; in this case the fifth, sixth and seventh, according to the most knowledgeable former practitioner.[72] In Halam, neither Arcturus (α Boötis) nor Spica (α Virginis) is used, even by a name other than Dhakarayn.

[72] Abdullah bin Mahan Al-Rashidi at the former stargazing point in Sinaw, 2 May 2006.

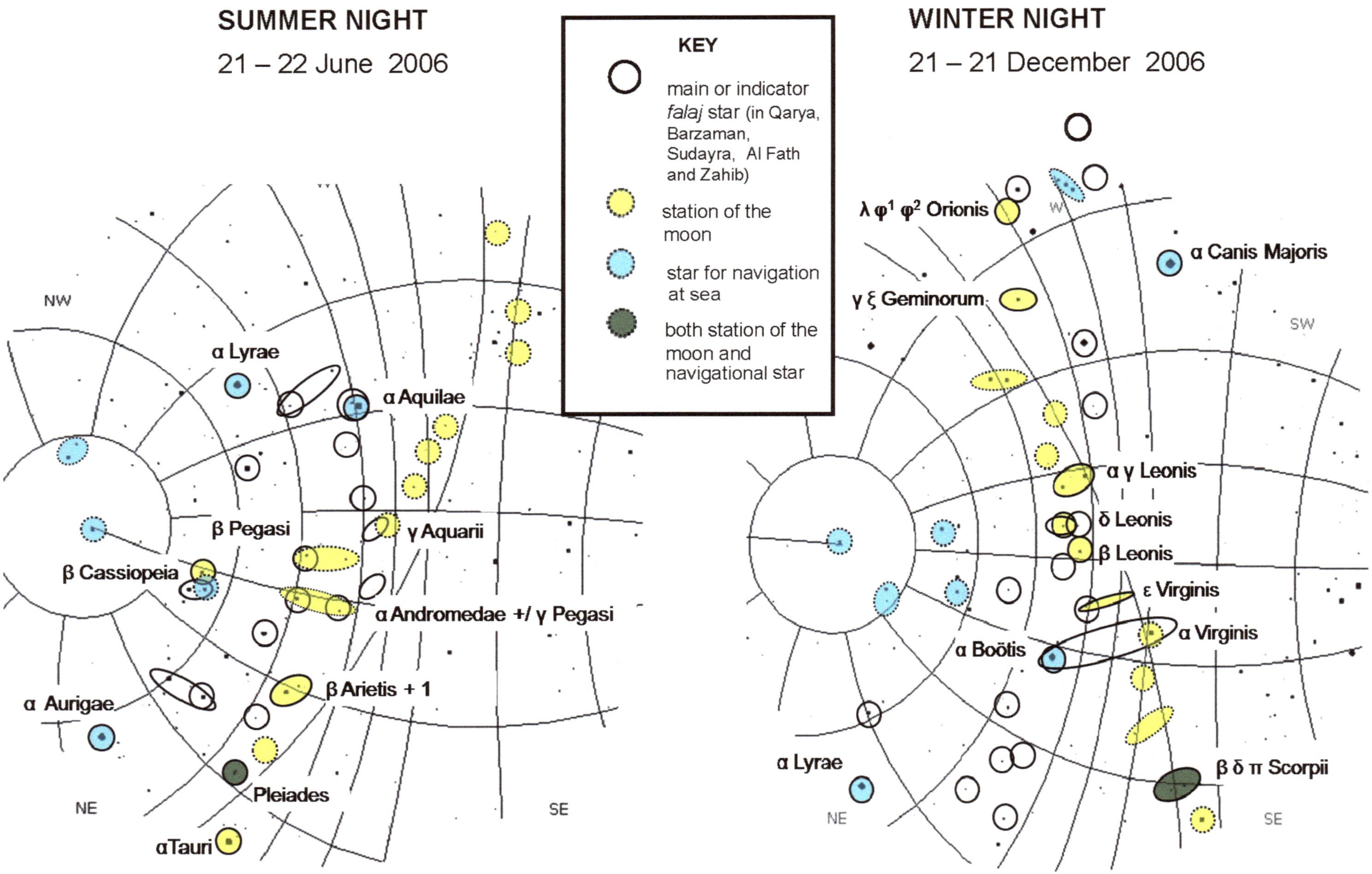

Figure 11.3 *Falaj* and navigation stars and stations of the moon

In Sudayra, Zahib and Al Fath one or two of the main stars are too faint to be seen most of the time under present day conditions of light and dust pollution. In Sudayra and Zahib, Thaqīla is a different faint star or star group. For both, alternative, brighter stars are apparently used frequently. 20 Canum Venaticorum, the brightest in the star cluster forming Thaqīla in Sudayra, has a magnitude of 4.71 while its substitute, Cor Caroli, α Canum Venaticorum, has a magnitude of 2.84 (the lower the magnitude, the brighter the star). The brightest star in Coma Berenices has a magnitude of 4.21, but this rises only ¼ hour before Vindemiatrix (Mīyathāb), insufficient for a space of two *athar*s (approximately one hour) and the true Thaqīla must be a fainter star that rises before this. The alternative brighter star used in Zahib was not identified, but is also in the area of Coma Berenices.

It is generally held that the naked human eye can see stars as faint as magnitude 6, but, based on the brightness of the stars identified in Qarya (Table 6.7), the faintest stars now regularly visible in the study villages, which are all affected to some extent by light pollution, have a magnitude of about 4.50. In any case, while many of the *falaj* stars are bright, this does not seem to have been a determining factor in their original selection.

As has been seen from both study areas (eg Tables 6.7 for Qarya and 9.3 for Al Fath), the time intervals between the rise of the main *falaj* stars is by no means equal. Taking only stars that command 2 *athar*s, the range is from 40 to 80 minutes in Qarya and from 40 minutes to 75 minutes in Al Fath. Given the number of stars in the sky, it should be possibly to find, say, 24 stars with a magnitude less than 4 at a time distance of close to one hour. However, it appears that the stars have been selected within a framework of well-known stars that are practically the same everywhere, i.e. many of those listed in Table 11.1 above. Since there is clearly a good deal of flexibility in the spaces between main stars, one wonders why certain bright and possibly better known stars, such as Sirius, were not more widely selected.

Navigational stars for the region and the stations of the moon are also shown on Figure 11.3. Approximately eight of the main *falaj* stars correspond with navigational stars. Given their very different uses, one for time keeping and the other to indicate points of the compass, there is no reason why even these eight should be the same. The one thing that they have in common is that they are all bright or readily recognised, but it cannot be concluded that there is any other relationship between them, especially given the difference in the names used for the two purposes. However, other stars were, and possibly still are, used for navigation on land and sea, and local names may differ to those in the literature. Some sea navigators in the region still use stars, and it should therefore be possible to find out much more about which stars are used and their local names. Only then will it be possible to assess any relationship with *falaj* stars.

In theory, the stations of the moon divide the night into roughly equal parts. However, from Figure 11.3 it can be seen that only 13 of the 28 stations of the moon coincide with a main *falaj* star, and if each village were considered separately the number would be less. The southerly stations appear to be less popular than northern ones for *falaj* use, possibly a function of visibility as stars far to the south are visible for shorter periods of the year than those in the east.

Of all these stars, only two are used for all three purposes: the Pleiades and Iklīl (in Scorpius). At present it appears that comparison of these different groups of stars takes us no closer to finding a reason for the selection of particular *falaj* stars.

The use of stars as indicators of weather and agricultural activities is widespread, not only in the Arabian Peninsula. The few indicators recorded during this study are all for *falaj* stars under their Omani names, indicating that the two are closely linked. Table 11.2 lists the sayings collected during this study. The dates given for Qarya are exactly the same as those given in the list of sayings thought to be from the Ministry of Agriculture (except for one or two typographical errors). That document has one or more sayings for each of 21 stars, some almost exactly the same as those collected in the study villages. Two examples are Thurayyā and Mūfī:

Thurayyā

Barzaman:

Lam tuṭla^c ath-thurayyā min ^cashayya: yazara^c al-burr hayya hayya.

When Thurayyā rises in the evening a lot of wheat is grown.

List possibly from Ministry of Agriculture:

اذا طلعت الثريا عشيا يارافق البر هيا هيا

Idhā ṭala^cat al-thurayyā ^cashayyā yārāfiq al-burr hayya hayya.

When Thurayyā rises in the evening it brings a lot of wheat.

Mūfī

Qarya:

يطلع الموفي تَجي جراب وتوفي

yaṭla^c al-mūfī tajī jurāb wa tūfī.

The rise of Mūfī brings bags full of dates and plenty (prosperity).

List possibly from Ministry of Agriculture:

اذا تنبت النخلة في الموفي تَجي جراب وتوفي

Idhā tanbat al-nakhla fī l-mūfī tajī jurāb wa tūfī.

If you cultivate date palms in Mūfī, it will bring you bags full of dates and plenty (prosperity).

It appears that there may once have been sayings for many of the main *falaj* stars, specific to agriculture in Oman, and that the sayings were similar in different places but are now largely forgotten. The information collected also shows that the sayings are not confined to the heliacal (dawn) rise of a star; for example, the heliacal rising of Ghurab is in winter, not summer.

Whether these sayings influenced the selection of stars for the *falaj* or vice versa, or whether they came into use at the same time is not known.

Star name from the study villages		Dawn rise*	Time for/of
Arabic	Transcription		
الكوكبين	Kawkabayn	20 April	
الفتح	Fatḥ		
الثريا	Thurayyā	16 May	much wheat
الدبران	Dubrān	1 June	dates, hot; wheat
الشابك	Shābik	17 June	
الظلمي / الجوزاء	Ẓalmī/Jawzā'	3 July	great heat
الشعرى	Sha^crā	19 July	plenty
الجنب	Janb	4 Aug	
الذراعين	Dharā^cayn	20 Aug	
البطين	Buṭayn	5 Sep	
السفاع	Sufā^c	21 Sep	agricultural season begins
المواثيب	Mawāthīb	6 Oct	garlic
الذكرين	Dhakarayn	22 Oct	summer and rain
الغفر	Ghafar	15 Nov	
العقرب/ الاكليل	^cAqrab/Iklīl	1 Dec	rain
كوي	Kuwī	21 Dec	planting date palms, seven days cold weather – animals die
الموفي/ النسر	Mūfī/Nisr	14 Jan	sugar cane juice; bags of dates
الغراب	Ghurāb	30 Jan	summer; bags of dates
الادم	Adam	15 Feb	no regret
الصارة	Ṣāra	3 Mar	juice
الصارة	Ṣāra	19 Mar	
السعد	Sa^cd	4 Apr	shearing; no cultivation
* Approximate date from Qarya			

Table 11.2 Agricultural sayings related to stars

11.6 *FALAJ* STARS AND THEIR NAMES

Stars are given different names depending on who is using them and for what purpose. Like creating pictures and stories from clouds in the sky, it needs only a little imagination and a few myths to think of descriptions of stars and their groupings. While many of the *falaj* star names are found in the literature on Arab star names others are not, and those that are may not be the same star as in the literature.

The variable spelling of star names and even of gender, for example the feminine adjective akhīra (meaning last) in Al Hamra compared with the masculine akhīr in Qarya for the second Ṣāra, is not surprising given that the information is transmitted orally. The gender of the Sha^carās is also variable, but given that in Arab mythology they are the two sisters of Suhayl (Badr, 1988: 156), one might expect them to be feminine.

The *falaj* stars all have names, but few others, even those used as dividers, are named. Canopus is known as Suhayl (α Carinae), and several of the seven stars of the Plough are named, even where they are not used for time keeping; the Pole Star, Polaris (α Ursae Minoris) is also known, as Jāh. Sirius (α Canis Majoris), is known as Sha^cara in most settlements; however, in Halam in Wadi Beni Jaber, near Sur, visited on 12-13 April 2008, Sirius does not have a name even though it is used to mark the path to Wadi Beni Khalid.

The similarity in star names in different villages in the Qarya area is notable. It is possible that one village learned from the other, with Qarya possibly being the first to use this system of stargazing. Although the variation in the endings of some names could have a significant effect on the meaning, most people do not think about the meanings of star names.

In the following sections, the available information of the *falaj* stars and their names is brought together, focussing on the main stars identified in this study. For comparison, reference is made to:

- star lists from the literature for other villages, even though many of these have not been identified;
- star names from other places collected during this study;
- stations of the moon (section 3.6), and
- navigational rhumbs (section 3.4).

Star names for villages not included as case studies in this work are given in Table 11.4. They come from various sources, some in English, others in Arabic and some in both and the sources are therefore given, in Table 11.3, with a description of the way names have been treated for inclusion in Table 11.4. Transliterations are provided so that both versions are available for comparison, in places making slight amendments to the original for consistency, but also taking into account the probable variation of pronunciation.

<table>
<tr><th colspan="2">Falaj star list</th><th rowspan="2">Source language</th><th rowspan="2">Treatment</th></tr>
<tr><th>Location</th><th>Source</th></tr>
<tr><td>Sual</td><td>Al-Siyabi (2000: 24-25)</td><td>Arabic and English</td><td>Arabic given as printed. English version modified to be consistent with the transliteration system used in this work.</td></tr>
<tr><td>Mudayrib</td><td>Al-Harthi c.1986</td><td>Arabic: Star chart shown in Figure 11.1</td><td>Arabic given as drawn.</td></tr>
<tr><td>Hageer</td><td>Al-Ghafri (2004: 72)</td><td>Arabic and English</td><td rowspan="3">Arabic given as printed except:
• the shadda for doubled letters is not given;
• the use of the letter /ذ/ in place of the /ض/ in the original for star names starting with Dhirāᶜ;
• the first two Ṣāras in Stall are given as صارة instead of شعرة.
The English transliteration differs to that printed and follows the system set out in the introduction, for example giving full weight to long vowels.</td></tr>
<tr><td>Stall</td><td>Al-Ghafri (2004: 71)</td><td>Arabic and English</td></tr>
<tr><td>Awabi</td><td>Al-Ghafri (2004: 70)</td><td>Arabic and English</td></tr>
<tr><td>Qarut</td><td>Wilkinson (1977: 110)</td><td>English</td><td>English given as printed except for endings with long vowels, which are given full weight to be consistent with the transliteration system used in this work;
Arabic version transliteration from the English.</td></tr>
<tr><td>Sumail</td><td>Al-ᶜAbrī (c. 1980: 42)</td><td>Arabic</td><td>Arabic given as printed. Transliteration to English taken from other lists.</td></tr>
</table>

Table 11.3 Written sources and treatment of *falaj* star names from the literature given in Table 11.4

<table>
<tr><th colspan="2">Izki area</th><th colspan="3">Wadi Beni Kharus</th><th>Sharqiya</th><th>Muscat area</th></tr>
<tr><th>Sumail</th><th>Qarut</th><th>Awabi</th><th>Stall</th><th>Hageer</th><th>Mudayrib</th><th>Sual</th></tr>
<tr><td>الكبكبين
Kabkabayn</td><td colspan="6">الكوكبين Kawkabayn</td></tr>
<tr><td colspan="2">الفتح Fatḥ</td><td colspan="3">الفتح Fateḥ</td><td colspan="2">الفتح Fatḥ</td></tr>
<tr><td>الثريا
Thurayyā</td><td>النجم
Najm</td><td colspan="3">الثريا Thurayyā</td><td>الثرياء
Thurayyā’</td><td>الثريا
Thurayyā</td></tr>
<tr><td colspan="2">بوقابل Bū Qābil</td><td>الدبران السفلي
Dubrān al-Suflī</td><td colspan="4">الدبران al-Dubrān</td></tr>
<tr><td colspan="2">الشابك Shābik</td><td>الدبران العلوي
Dubrān al-ᶜUlwī</td><td></td><td colspan="3">الشابك al-Shābik</td></tr>
<tr><td colspan="2">الظلمي Ẓalmī</td><td>الايمن Aymen</td><td colspan="2">اليمين Yamīn</td><td colspan="2">الظلمي Ẓalmī</td></tr>
<tr><td>الشعاير
Shᶜāyir</td><td>الشعرة
Shaᶜara</td><td>الشعرة البيضا
Shaᶜra l-Bayḍa</td><td>الشعاير
Shᶜāyir</td><td>الشعرة البيضا
Shaᶜra l-Bayḍa</td><td>الشعرء
Shaᶜrā’</td><td>الشعرى
Shiᶜrā</td></tr>
<tr><td></td><td></td><td>الشعرة الحمرا
Shaᶜra l-Ḥamrā</td><td></td><td>الشعرة الحمرا
Shaᶜra l-Ḥamrā</td><td></td><td></td></tr>
<tr><td>الجنب Janb</td><td>الجنب Janab</td><td>الجنب Ganab</td><td colspan="2">الجنب Ganb</td><td>الينب
Yanb</td><td>العبدين
ᶜAbdayn</td></tr>
<tr><td>name not given</td><td>الذراعين
Dhirāᶜayn</td><td colspan="2">الذراع Dhirāᶜ</td><td>الذراع الاول
Dhirāᶜ al-Awwal</td><td>الذراعين
Dhirāᶜayn</td><td>الرابعة
Rābiᶜa</td></tr>
<tr><td colspan="3">البطين Buṭayn</td><td></td><td>الذراع الاخير
Dhirāᶜ al-Akhīr</td><td>البطين
Buṭayn</td><td>السادسة
Sādisa</td></tr>
<tr><td>التقيلة
Thaqīla</td><td>الوقرت
Waqurt</td><td colspan="3">الفرفرة Farfara</td><td></td><td>السابعة
Sābiᶜa</td></tr>
<tr><td>المثاب
Muthāb</td><td>المثاب
Mithāb</td><td colspan="3">المواثيب Mawāthīb</td><td>المياثيب
Mīyāthīb</td><td>المثاب
Muthāb</td></tr>
<tr><td></td><td></td><td></td><td></td><td></td><td>الزبرة
Zubra</td><td></td></tr>
<tr><td colspan="7">Table 11.4 Star lists recorded in the literature from other villages in Oman cont[d] on next page</td></tr>
</table>

Izki area		Wadi Beni Kharus			Sharqiya	Muscat area
Sumail	**Qarut**	**Awabi**	**Stall**	**Hageer**	**Mudayrib**	**Sual**
الذكرين Dhakarayn		بو جبان Bū-Gabbān			الذكارة Dhakāra	الذكرين Dhakarayn
الغفر Ghafar	الغفر Ghufar	الغفر Ghafar		الغفر الاول Ghafar al-Awwal	الأغفر Aghfar	الغفر Ghafr
الزبان Zubān	الزبان Zabān	الزبانات Zabānāt		الغفر الاخير Ghafar al-Akhīr	الزبان Zubān	
القلب Qalb	غضير Ghaḍīr			العقرب ^{c}Aqrab	الاكليل Iklīl	
الكوي Kuwī	كوي الاول Kuwī al-Awwal	الكوي كوي Kuwī				الكوي Kuway
	كوي الثاني Kuwī al-Thānī	القواسم Qawāsim				
المنصف Munṣif	المنصف Munṣaf	المنصف Munṣif				
الموفي Mūfī		الطائر العلوي Ṭā'ir al-^{c}Ulwī	الطاير Ṭāyir		الطير Ṭayr	
		الطائر السفلي Ṭā'ir al-Suflī				الطائر Ṭā'ir
الغراب Ghurāb						
الأدم Adam	اللدم Ladam	الادم Adam				
الصارة الأولى Ṣara l-Ūlā	الصارة Ṣara	الصارة الاولى Ṣara l-Ūlā			الصارة Ṣāra	الصارة الاولى Ṣāra l-Ūlā
الصارة الثانية Ṣāra l-Thāniyya		الصارة الثانية Ṣāra l-Thāniyya	الصارة الوسطى Ṣāra l-Wusṭā		الصارة Ṣāra	الصارة الثانية Ṣāra l-Thāniyya
السعد Sa^{c}d		الصارة الثالثة Ṣāra l-Thālitha	الصارة الاخيرة Ṣāra l-Akhīra	الاخلع Akhla^{c}		السعد Sa^{c}d
				العرش ^{c}Arsh		

Table 11.4 Star lists recorded in the literature from other villages in Oman

Kuwī (كوي) = α Lyrae (Vega)
This star is used in all of the study villages and elsewhere in Oman. The name and the star appear to be the same everywhere, and in this form could mean "burning" or "blaze". In the literature, the usual name for α Lyrae is al-Nasr al-Wāqi^{c}, the perched eagle, but Paul Kunitzsch kindly pointed out that it is probably the same as Kuway (كوي), from the oldest *anwā'* traditions, in which it could refer either to α Lyrae (Vega) or α Aquilae (Altair).[73] The star name is given in this form in the list for Sual (Table 11.4), but the pronunciation there is not known. It is tempting to assume that the second Kuwī shown for Qarut by Wilkinson in Table 11.4 is Altair, but this cannot be done without at least knowing the method of watching or horizon used. As (al-Nasr) al-Wāqi^{c}, α Lyrae is one of the navigational rhumbs.

Manṣaf (المنصف)
Manṣaf means divider. This star is used in the Mudaybi area where, in the study villages, it is either or both of β Cygni (Albireo) and ζ Aquilae (Deneb el Okab).

Mūfī (الموفي)/Nis(a)r (النسر)
This star is called Mūfī, possibly meaning faithful or perfect, or Nisr (al-Nisr al-Ṭā'ir), the flying eagle. In the Qarya area it is Deneb, α Cygni. In the Mudaybi area it is α Aquilae (Altair) sometimes with γ Aquilae (Tarazed). Altair, under its standard Arabic name الطائر (al-Ṭā'ir), is also a navigational rhumb.

Ghurāb (الغراب) = α, β, δ + λ^2 Delphini
Ghurāb means the Crow. This is the same group of stars with the same name in all of the study villages and those in Table 11.4. It is not related to the international constellation Corvus (the Crow), also Ghurāb in standard Arabic. This is one of the few *falaj* stars that is not bright yet seems to be or have been widely used in Oman.

[73] Kunitzsch (1983: 116) and personal communication by letter dated 17 February 2008.

Adam/Ladam (اللادم الادم) = ε Pegasi (Enif)
Adam or Ladam is used in all of the study villages under the same name. No convincing dictionary definition of either form has been found.

The two Ṣāras (الصارة)
All of the study villages have two Ṣāras, with variable identities. Elsewhere there may be three, the third apparently in place of Sa^cd. A possible meaning for *ṣāra* is head or peak (eg of a mountain) (Lane, 1968. V4: 1744). In Qarya, the first Ṣāra (β Cassiopeiae, Caph) is the same star as the station of the moon named Baṭn al-Ḥūt, while the second Ṣāra in Qarya, Sudayra and Barzaman (α Andromedae, Alpheratz) is one of the two stars of the station called al-Fargh al-Thānī.

Sa^cd (السعد) = β Andromedae (Mirach)
There are many stars called Sa^cd, meaning good fortune or helping,[74] in Arabic astronomy but it appears that this is not one of them. It is the same star in all of the study villages.

Kawkabayn/Kabkabayn (الكوكبين) = β Arietis (Sheratan) + 1
Meaning the two stars, this is widely used for *falaj* purposes and, as its name suggests, comprises two stars: always (so far) β Arietis (Sheratan) with either α Arietis (Hamal) or γ Arietis (Mesarthim). As a station of the moon, all three stars (α, β, γ Arietis) are known as Sharaṭān.

Al-Fatḥ (الفتح)
Meaning the conqueror, this star is used in the Mudaybi area but not in the Qarya area. Its identification varies but 41 Arietis and β Persei (Algol) are popular.

Thurayyā (الثريا) = Messier object 45, the Pleiades
Meaning wealth, this star cluster is used in all the study villages under the same name, albeit with a variety of endings. In the Arabic literature it also has the name al-Najm, the star (eg Kunitzsch, 1961: 84). Thurayyā is both a station of the moon and a navigational rhumb. In Qarya it is also called Nathra (scattering), but not for *falaj* purposes.

Wilkinson gives a list of stars formerly used in Buraimi, which has not been included in Table 11.4 because it appears to be incomplete, with only 13 stars (1977: 110). In his list both Thurayyā and Najim appear, so clearly Najim was not the Pleiades in Buraimi. In Stall, Nijm is ε Virginis (Vindemiatrix), not Mawāthīb/Mīyathāb as elsewhere. This serves as a warning that the name alone is often not sufficient to identify a star.

Dabrān (الدبران) = α Tauri (Aldebaran)
Meaning the follower, this is another star used everywhere, sometimes with a different name: Bū Qābil in Barzaman, Muqbil in Sudayra also possibly meaning "the next", or following, and is a station of the moon. Sometimes the name is applied to two stars, the second being α Aurigae (Capella), as in Misfa and Barzaman. In Halam, in Wadi Beni Jabar, Dabrān is a cluster of about 10 stars, comprising the whole head of Taurus.

Shābik (الشابك)
Meaning the fishing net, Shābik varies from one place to another, as illustrated in Figure 11.4. Most frequently used (Qarya, Misfa, Sudayra, Zahib) are the same three stars (λ, φ^1 and φ^2 Orionis) that form Haq^ca, one of the stations of the moon. However in Al Fath, Shābik is γ Orionis and in Barzaman it is Orion's sword. In Stall, in Wadi Beni Kharus, Shābik is not a *falaj* star, but the name is used for Coma Berenices.

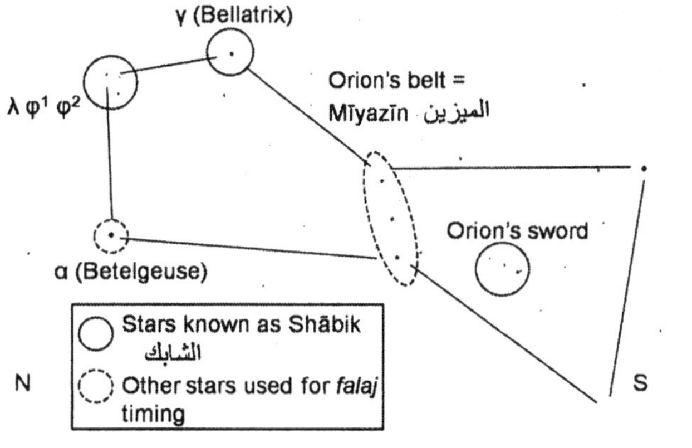

Figure 11.4 Shābik (fishing net) in Orion

Jawzā' (الجوزاء)/Ẓalmī (الظلمي) = γ Geminorum (Alhena)
With some variation in spelling, this is the same star in all the study villages with the possible addition of ϒ Geminorum (Alzirr) in Barzaman. I say possible because the second star in Barzaman is often an *^cayyuq* or "helper" rather than a main star and, given a flat horizon, ϒ Geminorum rises 13 minutes later than γ Geminorum; it is therefore more likely to be an *^cayyuq* than part of the main star. The name Jawzā', meaning "the couple" as in a man and a woman, is used in Arabic astronomy for the constellations of both Gemini and Orion.

The meaning of Ẓalmī is obscure: the dictionary definition of "dark" seems unlikely for a star. In Awabi, Stall and Hageer in Wadi Beni Kharus, the presence of a star called Yamīn/Aymen in this position does not indicate the use of a third name for the same star: in Stall, at least, Yamīn is Orion's belt (δ, ε + ζ Orionis), widely known elsewhere in Oman as Mīyazīn (the weighing scales; Figure 11.5). However, the navigational rhumb by the name of Jawzā' comprises Orion's belt, and does not include Alhena.

Gamma and ϒ Geminorum together form the station of the moon called al-Han^ca.

[74] Some of these may be named after pre-Islamic gods, eg Carey (2001: 275).

Sha^{c}ara (الشعرة) = α Canis Minoris (Procyon)/α Canis Majoris (Sirius)
Alpha Canis Minoris (Procyon) is used everywhere as a main *falaj* star, with some variation in speech/written form. In Al Fath and Zahib, Sirius is used as an indicator star; in Wadi Beni Kharus, where α Canis Majoris (Sirius) is used as a *falaj* star, Procyon is called Sha^{c}ra al-Ḥamra (red) while Sirius is Sha^{c}ra al-Bayḍa (white). The literary Arabic form for Procyon is al-Shi^{c}rā l-Ghumayṣā' (الشعرى الغميصاء), i.e. with eyes affected by weeping, while for Sirius it is al-Shi^{c}rā l-^{c}Abūr (الشعرى العبور), i.e. having crossed (the Milky Way), but the meaning of Shi^{c}rā itself is obscure. As Tīr, Sirius is a navigational rhumb.

Janb (الجنب) = 1 or more stars in the head of Hydra
Meaning the side or flank, this has the same name in all the study villages and many other places, usually comprising one or more of δ, ε and ρ Hydrae. ^{c}Abdayn (the two slaves) in this position in Sual (near the cement factory of Rusayl – in the Muscat area) is possibly a different star.

Dharā^{c}ayn (الذراعين) = α + γ Leonis (Regulus and Algeiba)
In MSA this is Dhirā^{c}ayn, the two arms, and as a *falaj* star always comprises α + γ Leonis. This is not related to the stars known as Dhirā^{c} in the literature, nor to the station of the moon called Dhirā^{c}. However, α Leonis is one of the stars in the station of the moon called Jabha.

Buṭayn (البطين) = δ Leonis (Zosma)
Perhaps more commonly pronounced Boṭayn and meaning the little belly, this star is used in all of the study villages, and in some others. So far it has always been found to be δ Leonis, and is not related to the station of the moon called Buṭayn (Table 5.3). This star is one of two in the station called Zubra, which again is different (based on the order of rise) to Zubra in Mudayrib.

Sufā^{c} (السفاع) = β Leonis (Denebola)
Possibly meaning "burning" or "a brand", this appears to be the same star in all of the villages in the Qarya area. Beta Leonis is also called Thaqīla in some villages in the Mudaybi area (see below), and under the name Ṣarfa, β Leonis is one of the stations of the moon.

Thaqīla (الثقيلة)
Meaning heavy, the identification of this star varies. However, it is β Leonis (Denebola) in Al Fath and Barzaman, in the Mudaybi area. Denebola is also used as a *falaj* star in Halam where it goes by the name of Farafra, possibly derived from *furfur,* meaning a small bird. Several places in Table 11.4 also have a Farafra in this position.

Mawāthīb/Mīyathāb (الموا ثيب/الميثاب)
Meaning leaping or the leapers, for *falaj* use this is generally one to three stars, with ε Virginis (Vindemiatrix) as the main one, γ Virginis (Porrima) the second and δ Virginis (Auva) the third. It is also known as a 7-star constellation in Qarya, which has been described in section 8.5. However in Stall, Mīyathāb comprises the four main stars in the international constellation Corvus, the Crow (δ γ β ε Corvi[75] - Algorab, Gienah, Minkar and Kraz), in Arabic Ghurāb (Figure 11.5).

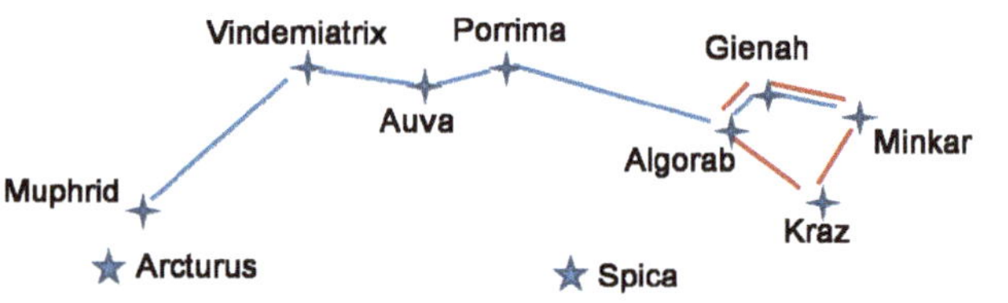

Figure 11.5 Mīyathāb (leaping) in Qarya and Stall
facing east; blue line for Qarya, red for Stall

In Halam, ε Virginis (Vindemiatrix) is known as Nijm, the star, not to be confused with the Pleiades, see Thurayyā above.

Dhakarayn (الذكرين) = α Boötis (Arcturus)
In Oman, Dhakarayn generally refers to two stars: α Boötis and α Virginis, but the second is rarely used for *falaj* purposes, and has not yet been found to be used on its own. In the literature these two stars are known as al-Simākayn, the two fishermen and α Virginis (Spica) forms the station of the moon called Simāk, while the navigational rhumb called Simāk is α Boötis. However, the name Simāk or Simākayn for the two is only known in Oman by people who have read about astronomy. Dhakar, meaning penis, is an old Arabic name for α Boötis,[76] but in Oman it is applied to both Arcturus and Spica. In Wadi Beni Kharus, the star in this position is called Bū Gabbān or Abū Gabbān (meaning n.k.), and I have identified this as α Boötis (Arcturus) in Stall.

Ghafr (الغفر) = α Coronae Borealis (Alphecca, Gemma)
Ghafr, Ghafar or Aghfar possibly meaning "the guard", has so far proved to be the same star in all villages.

Zubān (الزبان)
Zubān or Zibān, meaning claw of a scorpion, is used in the Mudaybi area and several of the villages in Table 11.4, but not in the Qarya area. In Sudayra, Al Fath and probably in Zahib, it is β Herculis (Kornephoros); in Barzaman it is probably γ Herculis. It is not related to the station of the moon called Zubānī or Zubānān, which comprises two stars in Libra (Table 5.3).

Iklīl (الاكليل)
Iklīl means crown or coronet, and it is used by this name only in the Mudaybi area (Sudayra, Zahib and Al Fath) and some other villages (Table 11.4), not in the Qarya area (see ^{c}Aqrab below). In Sudayra it comprises β and δ Scorpii (Graffias and Dschubba), in Zahib it is possibly δ

75 α Corvi is not as bright as the four stars mentioned here.
76 Kunitzsch (1961: 53) and personal communication letter dated 17 February 2008.

Herculis (Sarin) and in Al Fath it is α Herculis (Rasalgethi), which in Stall is called Zubān al-Thānī. The literary identification of the southern Iklīl is β, δ and π Scorpii, forming the station of the moon and navigational rhumb by this name, while the northern Iklīl is the Corona Borealis (see Ghafr above).

ᶜAqrab (العقرب)

ᶜAqrab is the scorpion and in the literature comprises at least part of the constellation Scorpius. However, for *falaj* purposes it is quite variable and of the study villages, the name is used only in the Qarya area, where it is a northern star, γ Draconis (Eltanin), though the southern equivalent, identified as β and δ Scorpii (equivalent to Iklīl above), is known. In Misfa and Al Hamra it appears that β, δ and π Scorpii were used, and the northern version may also have been known in Misfa.

While α Scorpii (Antares) does not form part of the constellation in the study villages, the use of the name Qalb (the heart) in this position in Sumail indicates that it may have been used elsewhere: Qalb (al- ᶜAqrab) is the literary Arabic name for this star. However, there are many *qalb*s/hearts in different constellations, so the Sumail star was not necessarily α Scorpii.

12 CONCLUSIONS

12.1 APPROACH TAKEN AND RESEARCH QUESTIONS

Timing water shares for an agreed distribution of irrigation water is fundamental to the operation of the traditional, community-managed *falaj* irrigation systems in Oman. The ready availability of wristwatches mean that there is now little technical reason to use the stars to time the allocation of water at night, and few of the communities with operational *aflāj* still use them. Since this ancient practice is rapidly disappearing, the primary aim of the research was to record the stars used and methods of stargazing, leading to the first and main research question: how do farmers use the stars to distribute water?

The small number of communities still using this ancient system leads to a second question: where and why are stars still used for this purpose?, putting the practice into its broader social and cultural context.

The conceptual framework sees the practice of stargazing as a means to avoid disputes over the division of water, and subsidiary questions are:
1) how are disputes over the allocation of water resolved? and, since disputes may be minimised by keeping records:
2) what written records are available regarding the allocation of water?

It was decided to concentrate on the area south of the Hajar Mountains, which is less affected by modernisation than the north. Five villages (Qarya near Al Hamra; Al Fath, Zahib, Sudayra and Barzaman all near Mudaybi) were found that still use stars, and here the methods of stargazing were investigated and the stars identified as far as possible. The management of their *falaj* systems was recorded in outline, as the framework within which stargazing is carried out. Information was also collected from a few other settlements no longer using the stars, in particular near Qarya, to provide a more complete case study.

It was not possible to provide a comparison with other countries in the region. It appears that the system of stargazing for agriculture in Oman is now unique. The closest country with similar irrigation systems is Iran, where the water clock was used for timing the allocation of water. Only recently has it become known that stars were used in parts of Iran until the 1970s for irrigation systems with a relatively high flow, but only where divisions of time shorter than 2 to 3 hours were not needed,[77] and there has not been an opportunity to obtain information on which stars were used and how.

[77] Majid Labbaf Khaneiki, Yazd International Qanat Center, personal communication, 5 May 2007.

12.2 HOW FARMERS USE THE STARS

The stars are used to tell the time at night. Whatever method is used for timing the principle is the same: when it is time for a person's share, they rearrange the sluices on the *falaj* channel so that the water flows to their fields. This study has identified a range of methods of stargazing, described in detail for the study villages in Chapters 6 to 10. It was found that the methods of watching and the stars used vary from place to place.

Factors considered in assessing the selection of method were:

- location;
- tribal complexity;
- learning/copying from others.

Location appears to be important, in that villages in or close to the mountains use the natural horizon, possibly because the horizon is relatively close and the rise or setting of stars is readily observed. Away from the hills the situation is more complex, and a range of methods was and is used. Based on the evidence collected, it could be argued that watching the natural horizon is the basis of the system, later developing (away from the mountains) into more complex but also, possibly, more accurate methods as the horizon became obscured, for example by urbanisation.

Tribal complexity does not appear to influence the choice of method: although villages with one main tribal group generally use the simpler methods, so also did settlements with a number of groups, such as Al Hamra and Qala. The situation when star use for the *falaj* first started may, of course, have been different, but although some of the information collected on water shares may relate to the situation 1000 years ago, as for the early water shares in Qarya, there is no such source for the method of stargazing.

Learning from or copying others is clearly an important factor for Zahib, which almost certainly took its method of stargazing from Mudaybi. In the Qarya area it appears that one settlement learned from another, with Qarya itself possibly being the first, but adapting the system to suit the local horizon.

A major achievement of this work is the identification of the majority of stars used in the study villages for the first time. They have the same names as those mentioned in agricultural folk lore but, as discussed in section 11.5, it is unclear how they were selected apart from generally rising in the east and setting in the west.

The names of the *falaj* stars often differ to those in modern Arabic astronomy, some coming from the earliest pre-Islamic traditions. Others may also come from the old traditions but were not recorded by the early writers. It appears that the stars were selected in pre-Islamic times, before any influence from Ptolemy, or the introduction of stations of the moon to the Islamic calendar. It was found that, because of the variation of star identifications from one place to another, it can be misleading to base the identification on name alone. No answer has been found as to why some places chose 24 stars and others only 21, nor as to why the spaces between them are not more equal in terms of modern time intervals.

12.3 WHERE AND WHY ARE STARS STILL USED?

Apart from the five settlements included in this work as case studies, only three others were known to still be using stars in 2008: Stall and Hageer in Wadi Beni Kharus, and Halam in Wadi Beni Jabar. Hence, of approximately 3000 *falaj* systems still in use in Oman only eight are known to still use the stars.

The question of why stars are still used is addressed first in the negative, i.e. the reasons for stopping using the stars, leading to conclusions as to why stars are still used. The traditional timing systems started to be abandoned when wristwatches became available, in some places before the accession of His Majesty, Sultan Qaboos, in 1970. However, the use of the sundial for the division of *falaj* water continues in many more places than does the use of stars. The fact that the use of the sundial is a daytime activity and most people prefer not to work at night should not be a major factor in this difference as irrigation is carried out 24 hours a day. Factors affecting the abandonment of *falaj* systems in general are given in Section 6.4. Other factors that specifically affect the decline of stargazing are:

- light pollution;
- the age (often old) and poor eyesight of people who know the stars and system of stargazing;
- the paucity of people prepared to learn the system, saying that it is too difficult.

In the five study villages, wristwatches are used when the skies are too cloudy to see the stars clearly, which happens more often with the increase of light pollution. Light (and dust) pollution is clearly a major factor in the rapid decline of the use of stars compared with the use of the sun. None of the *falaj* systems in towns visited during the research, including Adam, Bahla, Al-Hamra, Izki, Mintirib, Mudayrib, Mudaybi and Sinaw still use stars. For Sinaw and Sual, a village near Rusayl where a large cement factory was built in 1986-1987, this factor was given as the main reason for stopping the practice.[78]

The age of stargazers is also a factor, but a less direct one: more directly it is the reluctance of younger farmers to learn the system. In Qarya, it appears that stargazing will not continue after the current stargazer retires. In Sudayra also it is unlikely to survive another generation. In Barzaman, although not everyone uses the stars, there are some younger people and expatriate workers who know enough to carry out the water division, although it appears that only the two main stargazers, one now retired and the other with only one good eye, have a complete knowledge of the stars. The situation is similar in Zahib, with one or two people in their fifties having a thorough knowledge of the stars. Of the study villages, only in Al Fath does there appear to be a good chance of the system continuing, with young and old both using the system.

The conclusion must be that only where at least a few people have a thorough knowledge of the stars and time divisions, and everyone is prepared to use the stars, is their use likely to continue.

12.4 RESOLUTION OF DISPUTES AND RECORD KEEPING

Apart from Sudayra, where the traditional methods of *falaj* management have broken down altogether, none of the study villages admitted to having any disputes that could not be solved locally by the *falaj* administrator or village elders. Records are usually kept of financial transactions: the money raised for running the *falaj*, and payments for maintenance and repairs and to *falaj* officials. The ownership of water shares is not recorded in the study villages, apart from individual documents of transfer resulting from sales or inheritance, possibly because the ownership is well known and not disputed.

It is concluded that stargazing only continues where there are no significant disputes regarding water ownership.

12.5 SUMMARY

Traditional methods of timing *falaj* water using the sundial by day survives in many places, but the use of stars survives only in c. eight of the c. 3000 operational *falaj* systems. The continuation of such practices is seen as:

- adherence to tradition for its own sake;
- adherence to tradition to avoid disputes.

It is more difficult to explain why the eight villages still use stars. They are all places still highly dependent on agriculture, but the number of similar places could be in the hundreds, even taking into account the problem of light pollution. The answer must lie with the individual characteristics of each village, in particular the general lack of disputes over water ownership and time allocations, and in the personalities and wishes of those running the *falaj* system, usually the elders of the community.

Clearly the trend is for abandonment of traditional time keeping methods, but the communities deciding to do this occasionally regret the move. In Ridda, another satellite

[78] Abdullah bin Mahan Al-Rashidi, 2 May 2006; Hilal O. Siyabi, 30 January 2006.

town of Mudaybi, the use of stars ended in about 1997, but in c. 2006 the *falaj* community asked Saleh Al-Shidhani, an astronomer who comes from the village, to provide them with a schedule that replicates the former times determined by the stars. They would still use watches, but are not happy with modern equal time intervals, some feeling that they are losing out on the amount of water received.[79]

This study has succeeded in answering some of the research questions: the places where stars are still used, the stars themselves and the methods of watching them are now known. Information from the small number of places still using the stars provides a significant contribution to the knowledge of this traditional practice.

[79] Saleh Al-Shidhani, 4 November 2007.

APPENDICES

APPENDIX A STAR LISTS AND NAMES

A.1 Main stars used in Qarya

بسم الله الرحمن الرحيم

ت	النجم	من	الى	يوم	أثر
٠١	كوي ١	٣١/ديسمبر	١٣/يناير	٢٤	٣
٠٢	الموفي ١	١٤/يناير	٢٩/يناير	١٦	٢
٠٣	الغراب ١	٣٠/يناير	١٤/فبراير	١٦	٢
٠٤	الادم ١	١٥/فبراير	٢/مارس	١٦	٢
٠٥	الصاره (١) ١	٣/مارس	١٨/مارس	١٦	٢
٠٦	الصاره (٢) ١	١٩/مارس	٣/ابريل	١٦	٢
٠٧	السعد ١	٤/ابريل	١٩/ابريل	١٦	٢
٠٨	الكوكبين ٢	٢٠/ابريل	١٥/مايو	٢٤	٣
٠٩	الثريا(مجوعه) ٩	١٦/مايو	٣١/مايو	١٦	٢
١٠	الدبران ١	١/يونيو	١٦/يونيو	١٦	٢
١١	الشبكه(مجموعه) ٢	١٧/يونيو	٢/يونيو	١٦	٢
١٢	الجوزاء ١	٣/يوليو	١٨/يوليو	١٦	٢
١٣	الشعرى ١	١٩/يوليو	٣/اغسطس	١٦	٢
١٤	الجنب ١	٤/اغسطس	١٩/اغسطس	١٦	٢
١٥	الذراع ٢	٢٠/اغسطس	٤/سبتمبر	١٦	٢
١٦	البطين ١	٥/سبتمبر	٢٠/سبتمبر	١٦	٢
١٧	السفاع ١	٢١/سبتمبر	٥/اكتوبر	١٦	٢
١٨	المواثيب (مجموعه) ٧	٦/اكتوبر	٢١/اكتوبر	١٦	٢
١٩	الذكرين ٢	٢٢/اكتوبر	١٤/نوفمبر	٢٤	٣
٢٠	الغفر ١	١٥/نوفمبر	٣٠/نوفمبر	١٦	٢
٢١	العقرب (مجموعه) ٢	١/ديسمبر	٣٠/ديسمبر	٢٤	٣

#ـ كوي والموفي هما نجمان يؤخذ لهما في قوة البرد (الشتاء).
أما الشبكه والجوزاء فيؤخذ لهما في قوة الحر (الصيف) . ماجد بن خميس الصبحي

from Majid bin Khamis Al-Subhi, 9 November 2006

A.2 Modern schedule of stars formerly used in Al Hamra

مركز التنمية الزراعية بالحمراء

جدول بوضح عدد نجوم السنة وأوصافها ومواعيدها:

م	النجم	الوصف	الفترة		عدد	المدة	الزمن		ملاحظات
			من	الى	الاثار	باليوم	د	س	
١	الثرياء	مجموعة نجوم متقاربة	٥/١٢	٥/٢٧	٢	١٦	–	١	نجم صيفي
٢	الدبران	نجم واحد نير	٥/٢٨	٦/١٢	٢	١٦	٥	١	"
٣	الشبكة	مجموعة نجوم صغيرة متقاربة	٦/١٣	٦/٢٨	٢	١٦	٥٠	–	"
٤	الجوزاء	م واحد نير	٦/٢٩	٧/١٤	٢	١٦	١٥	١	"
٥	الشعرة	"	٧/١٥	٧/٣٠	٢	١٦	٢٠	١	"
٦	الجنب	مجموعة نجوم متقاربة	٧/٣١	٨/١٥	٢	١٦	٥٥	–	"
٧	الذراعين	نجمان متحدان	٨/١٦	٨/٣١	٢	١٦	٥٥	–	"
٨	البطين	نجم منفرد نير	٩/١	٩/١٦	٢	١٦	٤٠	–	"
٩	السفاع	"	٩/١٧	١٠/٢	٢	١٦	٢٠	١	"
١٠	الميثاب	ثلاثة نجوم معترضة	١٠/٣	١٠/١٨	٢	١٦	–	١	"
١١	الذكرين	نجمان نيران	١٠/١٩	١١/١١	٣	٢٤	٨	١	نجم شتوي
١٢	الغفر	نجم واحد نير	١١/١٢	١١/٢٥	٢	١٦	٣٠	١	"
١٣	العقرب	جملة نجوم معترضة	١١/٢٦	١٢/١٩	٣	٢٤	١٠	١	"
١٤	كوي	نجم نير	١٢/٢٠	١/١٢	٣	٢٤	٥٥	١	"
١٥	الموفي	نجمان نيران	١/١٣	١/٢٦	٢	١٦	٤٠	–	"
١٦	الغراب	مجموعة نجوم دقيقة	١/٢٧	٢/٩	٢	١٦	١٥	١	"
١٧	الأدم	نجم دقيق صغير الحجم	٢/١٠	٢/٢٣	٢	١٦	٤٥	–	"
١٨	الصارة الأولى	نجم متوسط الانارة	٢/٢٤	٣/١٠	٢	١٦	٥	١	نجوم بين الشتاء والصيف
١٩	الصارة الأخيرة	نجم أكثر إستنارة من الاولى	٣/١١	٣/٢٦	٢	١٦	٤٠	–	"
٢٠	السعد	نجم نير	٣/٢٧	٤/١١	٢	١٦	١٠	١	"
٢١	الكوكبين	نجمان نيران متقاربان	٤/١٢	٥/٥	٣	٢٤	٤٠	١	"

from Abdullah bin Muhammed bin Aamar Al-Abri 29 October 2007

A.3 Al Hamra stars from Muḥammad b. ᶜĀmar al-ᶜAbrī's version of Saᶜīd b. Sāliḥ al-ᶜAbrī's manuscript

Description checked ☺ successfully ☹ unsuccessfully; notes in italics: additional information from Badr al-ᶜAbrī's book (c. 1980: 32-41);

Abbreviations:
N = north, S = south, W = west and E = east; A = *athar* ; Cass = Cassiopeiae; BN = Banāt Naᶜsh (the seven stars in Ursa Major forming the Plough: α, β, δ + γ (al-Naᶜsh, meaning bier) and ε, ζ + η (al-Banāt, meaning daughters).

Name of *falaj* star	**Description + Location**	**Checks given in manuscript**	**Identification**	**Comment**
Dhakarayn (2 stars, 3 *athars*)	bright; some distance apart; N star is the indicator	with rise of star ½ A before, in same place as Dhakarayn, al-Shaᶜra *al-naᶜshiyya* (Procyon) in zenith ☺	N: Arcturus (α Boötis) S: Spica (α Virginis)	η Boötis (Muphrid) c. 19 min before Arcturus in same place
divider (1 star)	rises 1 *bāᶜ* to N of above after 1 A ☺	rises when Kawkabayn sets ☺ (for Hamal, α Arietis)	ε Boötis (Izar)	20 minutes from Arcturus
divider (1 star)	rises 1 *bāᶜ* to N of above, 2 A after al-Dhakarayn ☺		δ Boötis	40 minutes from Arcturus
Ghafr (1 star, 2 *athars*)	bright; rises in same place as 1st divider of Dhakarayn ☺	¾ A after: N star of Dhirāᶜān (γ Leonis) in the zenith ☹; 1 A after: setting of Thurayyā ☺.	Alphecca, Gemma (α Coronae Borealis)	Nearly 1 hr from α Coronae Borealis to γ Leonis in zenith.
2 stars	rise N of and 1$^1/_3$ A after Ghafr ☺; 1st 1 *bāᶜ* from Ghafr, 2nd 1 more *bāᶜ* to N; ☺		η + ζ Herculis	c. 50 min from α Coronae Borealis. Original identification of 2nd star as Rastaban incorrect: too far for 1 *bāᶜ*
ᶜAqrab (> 2 stars, 3 *athars*)	group; spread out; rises in SE, N of Suhayl (Canopus) ☺	rises as Dabarān sets ☺	Graffias, Dschubba (β + δ Scorpii)+ other(s) in Scorpius	Why not refer to exact number of stars?; OK for Aldebaran setting
divider (1 star)	bright; rises in same place as ᶜAqrab 1½ A later ☺	between 2 smaller stars ☺	Antares (α Scorpii)	lies between Al Niyat and τ Scorpii; rises in same location as π Scorpii
Kuwī (1 star, 3 *athars*)	bright; rises N of ᶜAqrab + its divider	rises as al-Shaᶜara *al-suhayliyya* (Sirius) sets ☺; ¼ A later, setting of the *ayyaq*, which precedes al-Dabarān ☺	Vega (α Lyrae)	Surprising that ᶜAqrab chosen as reference point being so far S. Definitely NOT Altair c. 10 min to setting of Capella.
divider (1 star)	smaller than Kuwī; rises N of Kuwī 1½ A later ☺	2 A after rise of Kuwī, setting of Procyon ☺	δ Cygni	1 hr from Vega: spot on for *athar* of 40 min. Procyon appears sets 80 min later.

Name of *falaj* star	Description + Location	Checks given in manuscript	Identification	Comment
Mūfī (2 stars, 2 *athars*)	bright; 1 rises in E (main marker), 1 where Kuwī rises ☺	preceded ¼ A by star rising in same place as N Mūfī ☺; at rise, star preceding Dhakarayn in the zenith ☺; 1 A after rise, Dhakarayn in the zenith ☺ 1 ¼ A from rise of E star, 1 less bright rising between the 2 main stars ☹;	E: Altair (α Aquilae) N: Deneb (α Cygni)	Sadr (γ Cygni) rises c. 12 min before a little S of Deneb, which rises in same place as Vega; Muphrid (η Boötis) fits with Arcturus for Dhakarayn. Gienah (ε Cygni) most likely but 4 min before time if *athar* 20 min long.
Ghurāb (>2 stars, 2 *athars*)	cluster of small stars; rises a little N of E Mūfī (Altair) ☺	at rise, 1st divider of Dhakarayn in zenith ☺	α, β, δ + λ2 Delphini	
divider (1 star)	smaller than Ghurāb; rises where al-Ghurāb rises ☺	rises1A after Ghurāb ☺; 2A after Ghurāb: star following Dhakarayn in the zenith ☺	1 Pegasi	good for time , location and magnitude: 4.06 compared w β Delphini's 3.62; OK for Ghafr, but why not mention by name?
Adam (1 star, 2 *athars*)	small; rises E of where Ghurāb rises ☺	When Ghafr in zenith, ¼ A before/from rise of Adam ☺	Enif (ε Pegasi)	location good relative to Ghurāb, Ghafr in zenith c 10 min after rise of Enif
divider of Adam (1 star)	a little smaller than Adam ☺; rises where Adam rises	1 A after Adam: rise of divider; 1st star of Dhirāᶜān (Regulus) sets ☹	Baham (θ Pegasi)	rises a little S of where Enif rises, but fits the bill: apparent magnitude 3.50 compared to Enif's 2.37. Regulus sets c 10 min before.
Ṣāra l-Ūlā (1 star, 2 *athars*)	middling bright; rises N of Ghafr (not a useful description as the two stars are far apart in time)	½ A before Ṣāra, a bright star rises at same point; another star rises at the same time W of BN ☺; a star follows after ½ A - 2 *bāᶜ* E of Ṣāra ☺; 1¼ A after Ṣāra: 2 bright stars, 1st rising in same place as that following Ṣāra 1 ☺	Scheat (β Pegasi)	probably Matar (η Pegasi), 24 min before, but not v bright; Caph (β Cass) Markab (α Pegasi) Schedar (α Cass) + γ Cass. No info. on divider for Ṣāra 1.
Ṣāra l-Thāniyya (1 star, 2 *athars*)	brighter than and rises a little N of Ṣāra 1☺		Alpheratz (α Andromedae)	magnitude 2.06 compared with Scheat's 2.43
divider (1 star)	smaller than + rises at same place as Ṣāra 2 after 1A ☺	followed by 1 rising in the E in same place as (the divider) for Ṣāra 1, ¾ A after Ṣāra 2	δ Andromedae	possibly Ruchbah (δ Cass.)

Name of *falaj* star	Description + Location	Checks given in manuscript	Identification	Comment
Sa^{c}d (1 star, 2 *athars*)	bright; rises 1 *bā^c* by eye N of Ṣāra l-Thāniyya ☺	At rise, Kuwī in the zenith ☺; S star of Dhakarayn ☺ + 1st star of BN ☹ set	Mirach (β Andromedae)	1st star of BN sets before Spica on Qarya's photorealistic horizon.
1 star	1 *bā^c* N of above after 1¼ A; similar brightness ☺		Almach (γ Andromedae)	1¼ A = 42 min.
Kawkabayn (2 stars, 3 *athars*)	close (S+N); N is brighter; rises in E ☺	¼ A before: bright star 1 *dhirā^c* to N ☺; 4th star of BN sets close to rise of main star; ¼ A or more after: S Mūfī in zenith 1¼ A after: 2 stars rise, 1 in same place as Kuwī ☺, other as *^{c}ayyaq* al-Dabarān; 1½ A after: 1 small star rising from main star in same place as Kawkabayn ☺; 1½ A after: N Dhakarayn (Arcturus) sets ☺ ; 2¼ A after: 7th star of BN sets ☺	Sheratan (β Arietis) + Hamal (α Arietis - N) or Mesarthim (γ Arietis - S)	Rasalmothallah (α Trianguli) – 13 minutes before Algol (β Persii) + κ Persii/Mirfak (α Persii); κ Persii closer to Capella; 41 Arietis fits well with time of rise and location (a little N of Hamal) Spot on. Time fits.
Thurayyā (cluster, 2 *athars*)	cluster close together; looks like a bunch of grapes; rises at near + N of Kawkabayn	star middling bright rises ¼ A before in same place as Kuwī ☺; ¾ A after Thurayyā, middling bright rising in same place, looks similar to above ☹; Adam in zenith ½ A after main star ☺	Pleiades	ε Persii fits, magnitude 2.87, 10 min before 37 Tauri fits time + rises close to Pleiades, but magnitude only 4.34; no obvious contender rising near ε Persii;
divider (1 star)	rises S of Thurayyā ☺ 1 A after.	Ghafr sets 1 A + a bit after main star rises ☺	λ Tauri	
^{c}Ayyaq (1 star)	brighter/bigger than Dabarān; rises N (S) of Kuwī ☺	¼ A before Dabarān	Capella (α Aurigae)	
Dabarān (1 star, 2 *athars*)	bright; (1 *bā^c* S of Thurayyā)		Aldebaran (α Tauri)	as it rises S of Pleiades cannot be Capella
divider (1 star)	close to Dabarān in brightness; rises between ^{c}Ayyaq and Dabarān1A after ☺	then star rises in same place as ^{c}Ayyaq middling bright, 1 ¼ A after main star ☺	El Nath (ζ Tauri)	Menkalinan (β Aurigae)
Shubka (>2 stars, 2 *athars*)	close group of small stars; rise in E 1 *bā^c* S of Dabarān ☺	preceded by a bright star 1 *dhirā^c* to S by ¼ A ☺;	Meissa (λ Orionis) + others	Bellatrix (γ Orionis) precedes
divider (1 star)	bright, rises in same place ☺	1 A after Shubka ☺	Betelgeuse (α Orionis)	good for 20 min *athar*

Name of *falaj* star	Description + Location	Checks given in manuscript	Identification	Comment
Jawza (1 star, 2 *athars*)	bright; rises in same place as Pleiades ☹		Alhena or Mebsuta (γ or ε Geminorum)	Alhena bright but rise further; Mebsuta location good but not v bright (3.03)
divider (1 star)	bright; rises after 1A in same place as divider of Dabarān (ζ Tauri) ☺	1½ A after Jawza another similarly bright star rises in the same place called Dhirā^c at same time as rise of Sirius ☺; 1½ A after Jawza, Kuwī sets ☺	Castor (α Geminorum)	Pollux (β Geminorum)
Sha^{c}ra (1 star, 2 *athars*)	the N one (*na^{c}shiyya*) bright; rises E of Shubka	When Kawakabayn in zenith, ¼ A or less from rise of main star ☺	Procyon (α Canis Minoris)	
divider (1 star)	smaller/less bright than Procyon; rises 1 *bā^c* N of Procyon ☺		Altarf (β Cancri)	a short *bā^c* but no alternative *rises S of Procyon in book misleading*
Janb (> 2 stars, 2 *athars*)	cluster ☺; rises a little NE of Procyon	Suhayl rises ½ A later (☺ for 1st star), Thurayyā in zenith 1½ A later ☺	head of Hydra	about same place as rise of Procyon, but has to be S of Dhirā^{c}ān, so probably OK
divider (1 star)	brighter than Janb; rises in same place as divider Dabarān (El Nath) after 1A	1st star of BN (Dubhe) rises approx ¼ A before; 2nd (Merak) rises ½ A later ☺	? Rasalas or Ras Elased Australis (μ or ε Leonis)	1st better location, 2nd better timing
Dhirā^{c}ān (2 stars, 2 *athars*)	bright; rise at same time; 1 *bā^c* apart; rise 1 *bā^c* NE of Janb ☺ (for Regulus)	2 fainter stars to N rise ¼ A before ☺; 3rd star of BN rises 1½ A later ☺; Dabarān in zenith 1 A later ☺	Algeiba + Regulus (γ + α Leonis)	Aldhafera (ζ Leonis)+ η Leonis
divider (1 star)	smaller than the above; rises in same place as S of above	1A after main star	*ρ Leonis*	rises a bit S of Regulus
Baṭn (1 star, 2 *athars*)	bright: rises in same place as Dhirā^{c}ān ☺ (for Algeiba)	4th star from BN rises ¼ A before ☺; bright star S of Mīyazīn in the zenith, ¾ A from rise of Baṭn; star before Shubka in zenith after 1¼ A	Zosma (δ Leonis)	probably Rigel (β Orionis) ? Bellatrix (γ Orionis)
Sufā^c (2 stars, 2 *athars*)	bright; rises 1 *bā^c* S of Baṭn, E of Thurayyā	preceded by ¼ A by 5th star of BN; divider of Shubka in zenith ¼ A after rise ☺; 6th star in BN: ¾ A after rise of main star ☺	Denebola (β Leonis)	*4th BN given in book not possible*
divider (1 star)	rises in same place as Kuwī after 1A	rise of 7th star from BN means ¼ A left from time of Sufā^c ☺	Cor Caroli (α Canum Venaticorum)	
Mawāthīb (>2 stars, 2 *athars*)	middling bright; spread out in a line; rise in same place as Sufā^c small difference in time of rise; a bit S of E	7th from BN ¼ A before ☺; Sirius in zenith close to rise.	Vindemiatrix, Porrima + probably Auva (ε, γ + δ) Virginis	only 1 can rise in same place as single star =Vindemiatrix, the furthest north

Name of *falaj* star	Description + Location	Checks given in manuscript	Identification	Comment
divider (1 star)	rises 1 A before and in same place as Dhakarayn	at rise of this star, star before Procyon in zenith		no obvious star: unclear which Dhakarayn meant.
1 star	brighter; at same place as N star of Mawāthīb, 1.5 A (45 min)	1½ A from Mawāthīb, Procyon in zenith ☺ rise of star preceding N Dhakarayn after 1½ A ☺	Muphrid (η Boötis)	OK on time + magnitude: Muphrid 2.64 compared to 2.84 for Vindemiatrix.

Note: times of rise, setting and stars in the zenith were checked on Starry Night™. A photorealistic horizon was prepared for Qarya and used in checking the timings. Some variation is therefore to be expected since the horizons watched in Al Hamra would have been different.

Checks on stars in Al Hamra by distances given in the manuscript

Stars (mainly 1 *bāᶜ* apart)	Declension at rise (N)	Difference	Comment
Arcturus	19° 8.2'		
Izar	27° 2.2'	7° 54'	
δ Boötis	33° 16.9'	6° 15'	
Alphecca	26° 41.1'		
ζ Herculis	31° 35'	5°	
η Herculis	38° 55'	7°	
Scheat	28° 7.7'		
Markab	15° 15'	12° 51'	long for 1 *bāᶜ*
Alpheratz	29° 5'		
Mirach	35° 40'	6° 35'	
Mirach	35° 40'		
Almach	42° 22'	13° 18'	long for 1 *bāᶜ*
Rasalmothallah	29° 37'		
Hamal	23° 30'	6° 07'	
Sheratan	20° 51'	8° 14'	
Pleiades	24° 13'		
Aldebaran	16° 32'	8° 31'	
Elnath	28° 37'		
ζ Tauri	21° 9'	7° 28'	
Meissa	9° 56'		
Bellatrix	6° 21	3° 35'	this is 1 *dhirāᶜ*
Procyon	5° 12'		
Altarf	9° 09'	3° 57'	? short for 1 *bāᶜ*
Algeiba	19° 48'		
Regulus	11° 56'	8°	
Al Janb (δ Hydrae)	5° 40'		
Regulus	11° 56'	6°16'	
Zosma	20° 29'		
Denebola	14° 31'	6°	

A.4 Main stars formerly used in Qala

حساب النجوم

م	اسم النجم	تاريخ الدخول	المدة	ملاحظة
١	المولي	١/١٥	١٥ يوم	٤٠ مربعي (قوة البرد)
٢	الغراب	١/٣٠	١٦ يوم	
٣	الادم	٢/١٦	١٦ يوم	
٤	الصارة الأولى	٣/٢	١٦ يوم	
٥	الصارة ألا خير	٣/١٧	١٦ يوم	
٦	السعد	٤ /٣	١٦ يوم	
٧	الكوكين	٤ /١٨	٢٤ يوم	
٨	الثرية	٥ /١٢	١٦ يوم	
٩	الدبران	٥/ ٢٨	١٥ يوم	
١٠	الشبكة	٦/١٢	١٦ يوم	رجوع الشمس وتنتقل لشدة الصيف ٦/٢٢
١١	الجوزه	٦/٢٦	١٦ يوم	
١٢	الشعرة	٧/١٢	١٥ يوم	
١٣	الجنب	٧/٢٦	١٦ يوم	
١٤	الذراعين	٨/١٢	١٦ يوم	
١٥	لبطن	٨/٢٨	١٦ يوم	
١٦	السفاع	٩/١٣	١٦ يوم	
١٧	الميثاب	٩/٢٨	١٦ يوم	
		١٠/١٤	١٠ يوم	تخطي لا يكون فيه حساب نجوم
١٨	الذكرين	١٠/٢٤	٢٣ يوم	
١٩	الغفر	١١/١٦	١٥ يوم	رجوع الشمس وتنتقل لشدة الشتاء ١٢/٢١
٢٠	العقرب	١٢/١	٢٣ يوم	
٢١	كوي	١٢/٢٣	٢٣ يوم	

from Harith bin Nabhan Al-Abri, November 2006

APPENDIX B
Use of the term *qaḍiyān* in Mudaybi

Saeed bin Aamar bin Saud Al-Harthi, in charge of organising the order of water distribution for the falaj in Mudaybi, gave the following explanation of the use of the term *qaḍiyān* for *falaj* water shares.[°]

The day is divided into 4 quarters:

qaḍiyān: sunset to midnight

min al-layl (from the night): middle of the night to sunrise

qaḍiyān: sunrise to midday

min al-nahār (from day time): midday to sunset.

Each quarter has 12 *athar*s.

The first 11 *athar*s are named after the section, i.e. 1 - 11 *al-qaḍiyān*.
The 12th is called by the name of the division for the following section, so the 12th *athar* after sunset is actually called *niṣf al-layl* (half the night).

It appears that the use of this term always applies to ¼ of the 24-hour day, but otherwise varies among villages. The main point is that it explains why less than 12 *athar*s actually take the name of the division.

Saud bin Abdullah Al-Habsi's (*wakīl* of the *aflāj* in Zahib and Mudaybi) grandfather wrote a book 21 pages long on the water rights of the *falaj* of Mudaybi in 1335/1917. The first page of this is given below: it is in a standard form, stating the amount of water (in *qaḍiyān*) that each person owns. The document is all in one hand, with a few corrections in the same hand. It is in the standard form, stating the amount of water that each person owns. The second entry, for example, says:

Li haq qaḍiyān thalāth wa niṣf qaḍiyān arbaᶜ li Ḥamad bin Rāshid

meaning that Ḥamad bin Rāshid owned the third and half of the fourth *qaḍiyān*.

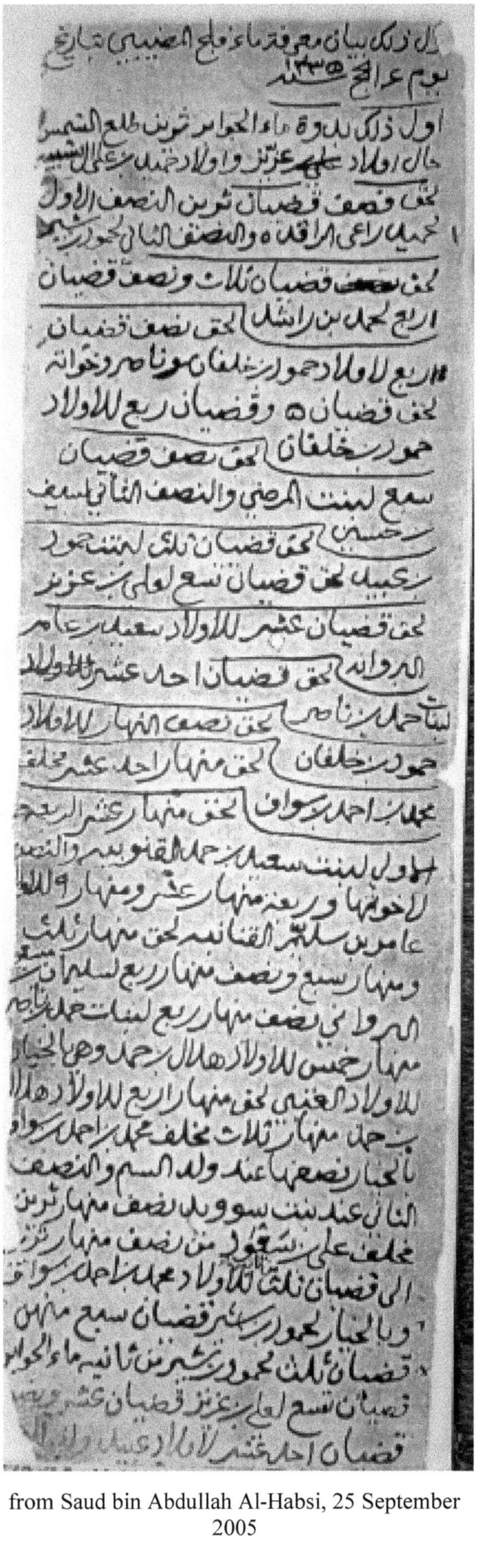

from Saud bin Abdullah Al-Habsi, 25 September 2005

° Interviewed on 17 April 2008.

APPENDIX C

Months of the Islamic Year

The Islamic (Hijrī) year has 12 months each 29 or 30 days long, starting with Muḥarram and ending with Dhū l-Ḥijja. The Hijrī months are listed below:

Arabic	Transliteration
محرم	Muḥarram
صفر	Ṣafar
ربيع الاول	Rabī^c al-Awwal
ربيع الثاني	Rabī^c al-Thānī
جمادى الاولى	Jumādā l-Ūlā
جمادى الاخرة	Jumādā l-Ukhra
رجب	Rajab
شعبان	Sha^{c}bān
رمضان	Ramaḍān
شوال	Shawwāl
ذو القعد	Dhū l-Qa^{c}da
ذو الحجة	Dhū l-Ḥijja

Glossary of *falaj* terms used in this study

Falaj term		Meaning
Arabic	**Transcription**	
أفلاج	*aflāj*	Plural of *falaj*. Omani irrigation systems, which supply water for a community of farmers.
عين	*ᶜayn*	Spring. There are many springs from the limestone in the Hajar Mountains
عيني فلج	*ᶜaynī falaj*	*Falaj* whose water comes from a spring "*ᶜayn*".
عريف	*ᶜarīf*	Officer/foreman of the *falaj*. He is usually in charge of verifying the daily water distribution and/or maintenance of the *falaj* tunnel or canals.
أثر	*athar*	The basic unit of water share in most of the *aflāj* of Oman. Theoretically it is equal, on average over the year, to 30 minutes.
بادة /بدة	*badda/bādda/bāda*	Unit for distributing *falaj* water. The 24-hour day is usually divided into a night-time *badda* and daytime *badda*, each with 24 *athars*. In some *aflāj*, the day is divided to three *baddas* or in few cases the *badda* is equal one full day, but there are still 24 *athars* by day and 24 by night.
دلال	*dallāl/dillāl*	Auctioneer. He organises and sometimes documents the regular auctions of water.
داؤودي فلج	*dā'ūdī falaj*	This type of *falaj* taps groundwater by means of a tunnel.
دوران	*dawrān*	The irrigation rotation, i.e. the time needed to irrigate the cultivated area.
فلج	*falaj*	Singular of *aflāj*.
غيل	*ghayl*	Base flow in the upstream section of a wadi.
غيلي فلج	*ghaylī falaj*	*Falaj* using *ghayl* water.
قاعدة	*qāᶜida*	A water share allocated for *falaj* maintenance. This water is auctioned or rented periodically.
محاضرة	*muḥāḍara*	The process of time reckoning for the irrigation schedule; the place where this is carried out; the system of timing the allocation of water.
مزيدة	*muzayyida*	A water share allocated for *falaj* maintenance, as for the *qāᶜida*.
مغسلة	*mughsila*	Place on *falaj* for washing corpses.
ردة	*radda*	A share of *falaj* water, often equivalent to the *badda*.
سهم	*sahm*	A share of *falaj* water equivalent to 1½ *athars*, i.e. nominally ¾ hour.
شريعة	*sharīᶜa*	Place where the main stream of the *falaj* is first utilized for domestic purposes, not Islamic law.
طاسة	*ṭāsa*	A water clock for timing water shares.
وكيل	*wakīl*	The *falaj* administrator. He is usually in charge of the overall administration of the *falaj*, including water distribution, servicing and managing the *falaj* income.
وقف	*waqf*	Property of mosques used for running mosques, schools and for charity.

BIBLIOGRAPHY

Abercrombie, Thomas J. 1981
"Oman: guardian of the Gulf", *National Geographic*, v. 160: 344-377.

al-ᶜAbrī, Badr, c. 1980
Al-bayān fī baᶜḍ aflāj ᶜUmān. (A Description of some of Oman's *Aflāj*): Muscat.

Agius, Dionisius A. 2005
Seafaring in the Arabian Gulf and Oman: People of the Dhow. London/New York/Bahrain: Kegan Paul.

Agius, Dionisius A. 2008
Classic Ships of Islam from Mesopotamia to the Indian Ocean. Leiden/Boston: Brill.

Ali, Yusuf Abdullah 1968
The Holy Quran: text, translation and commentary. Beirut: Dār al-ᶜArabiyya.

Alian, Mohamed A. 2006
In Memory of the Sambuq, a record of a way of life. Muscat: Historical Association of Oman.

Arunachalam, B. 1996
"Traditional sea and sky wisdom of Indian seamen and their practical applications", *in* H. P. Ray, and J.-F. Salles, eds., *Tradition and Archaeology: Early Maritime Contacts in the Indian Ocean.* New Delhi/Lyon: National Institute of Science, Technology and Development Studies & Maison de l'Orient Méditerranéen: 261-281.

Badger, George Percy 1881
An English-Arabic Lexicon: London, C. Kegan Paul & Co..

Badr, ᶜAbd al-Raḥīm A. M. 1988
Mawsūᶜat asmā' al-nujūm ᶜinda l-ᶜarab fī l-falk al-qadīm wa l-ḥadīth (Encyclopaedia of Arab Star Names in Ancient and Modern Astronomy): Amman.

al-Baṭāshī, Seif Ḥ. Ḥ. 1998
'Itḥāf al-ᶜayān fī tārīkh baᶜda ᶜulumā' ᶜUmān (Presentation of the Views on History by some of the *ᶜUlumā'* of Oman): Muscat, Maktabat al-Mustishār.

Beckingham, C.F and R. B. Serjeant 1950
"A Journey by Two Jesuits from Dhufār to San'ā in 1590" *Geographical Journal*, v. 115, No. 4/6: 194-207

Beaumont, Peter, Michael E. Bonine and Keith S. MacLachlan, eds. 1989
Qanat, Kariz and Khattara: Traditional Water Systems in the Middle East and North Africa. Cambridgeshire: Middle East and North Africa Studies Press Ltd.

Ben Shemesh, Aharon 1958
Taxation in Islam. Volume I: Leiden: E.J. Brill.

Ben Shemesh, Aharon 1965
Taxation in Islam. Volume II: Leiden/London: E.J. Brill and Luzac & Co. Ltd.

Ben Shemesh, Aharon 1969
Taxation in Islam. Volume III: Leiden/London: E.J. Brill and Luzac & Co. Ltd.

Bonine, Michael E. 1996
"Qanats and rural societies: sustainable *agriculture and irrigation cultures in* contemporary Iran", *in* J.B. Mabry, ed., *Canals and Communities: Small-scale Irrigation Systems.* Tucson: University of Arizona Press: 183-209.

Carey, M. 2001
Painting the stars in a century of change : a thirteenth-century copy of al-Ṣūfī's Treatise on the fixed stars. PhD Thesis, University of London.

Cleuziou, Serge and Maurizio Tosi 2007
In the Shadow of the Ancestors. The Prehistoric Foundations of the Early Arabian Civilisation in Oman. Muscat: Ministry of Heritage and Culture.

DeYoung, Gregg 2000
"Astronomy in Ancient Egypt", *in* H. Selin, ed., *Astronomy Across Cultures: The History of Non-Western Astronomy*: Dordrecht/Boston/London: Kluwer Academic Publishers: 475-508.

Donaldson, William J. 1978
"Fishing and Fish Marketing", *Research and Development Surveys in Northern Oman.* Volume V, University of Durham.

Dutton, Roderick W. 1995
"Towards a secure future for the aflaj in Oman" *Water Resources Management in Arid Countries.* Muscat: 16-24.

Encyclopaedia of Islam. 2006
New Edition. Index Volume Fascicule 2 Glossary and Index of Terms. Leiden/Boston: Brill

Evans, James E. 1998
The History and Practice of Ancient Astronomy. New York: Oxford University Press.

Al-Ghafri, Abdullah S. 2004
Study on Water Distribution Management of Aflaj Irrigation Systems of Oman. PhD thesis, Hokkaido University.

Goblot, Henri 1979.
Les Qanats – Une Technique d'Acquisition de l'Eau. Paris/La Haye/New York: Mouton Editeur.

al-Ḥawārī, Mohammed b. 1975
Jāmiᶜ Abī l-Ḥawārī (Collection of Abū l-Ḥawārī). Muscat: Ministry of National Heritage and Culture.

Hesiod 1973
Theogony; Works and days / Hesiod. Elegies / Theognis. Translated and with introduction by Dorothea Wender. Harmondsworth: Penguin.

Jorgensen, Donald G. and Walid Y. Al-Tikriti 2002
"A hydrologic and archeologic study of climate change in Al Ain, United Arab Emirates", *Global and Planetary Change*, v. 35: 37-49.

Kak, Subhash 2000
"Birth and early development of Indian astronomy", *in* H. Selin, ed., *Astronomy Across Cultures: The History of Non-Western Astronomy*: Dordrecht/Boston/London: Kluwer Academic Publishers: 303-340.

al-Kharūṣī, Muhannā b. Khalfān b. ᶜUthmān 1465/2004
Al-nujūm al-zāhira fī l-aflāk al-rā'ida (Bright Stars in the Celestial Spheres): Muscat: Maktabat al-Mustishār.

al-Kindī, Abū Bakr Aḥmad b. ᶜAbdallāh b. Mūsā 1983
Al-Muṣannaf (The Compilation). Part 17. Muscat: Ministry of National Heritage and Culture.

King, David A. 2000
"Mathematical astronomy in Islamic civilisation", *in* H. Selin, ed., *Astronomy Across Cultures. The History of Non-Western Astronomy.* Dordrecht/Boston/London: Kluwer Academic Publishers: 585-613.

Kunitzsch, Paul 1961
Untersuchungen zur Sternomenklatur der Araber. Wiesbaden: Otto Harrassowitz.

Kunitzsch, Paul 1983
Über eine anwā'-Tradition mit bisher unbekannten Sternnamen. Munich: Bavarian Academy of Science and C.H. Beck Verlag.

Kunitzsch, Paul 1989 a
The Arabs and the Stars: Texts and Traditions on the Fixed Stars, and their Influence in Medieval Europe. Northampton: Ashgate Publishing Group.

Kunitzsch, Paul 1989 b
"The Astronomer Abu l-Ḥusayn al-Ṣūfī", *The Arabs and the Stars: Texts and Traditions on the Fixed Stars, and their Influence in Medieval Europe.* Part XI. Northampton: Ashgate Publishing Group.

Kunitzsch, Paul 2004
Stars and Numbers: Astronomy and Mathematics in the Medieval Arab and Western Worlds. Northampton: Ashgate Publishing Group.

Lane, Edward W. 1968
An Arabic-English Lexicon. Volumes 1 to 8. Beirut: Librarie du Liban.

Letts, Sally E. 1978
Groundwater and its Present Day Use in Part of Northern Oman. PhD thesis, University of Oxford.

Lightfoot, Dale R. 2000
"The origin and diffusion of qanats in Arabia: new evidence from the northern and southern peninsula", *The Geographical Journal*, v. 166: 215-226.

Maktari, Abdulla M. A. 1971
Water Rights and Irrigation Practice in Laḥj. Cambridge: Cambridge University Press.

Menzel, Donald H. 1975
A Field Guide to the Stars & Planets. London: Collins.

Ministry of Information, Sultanate of Oman 2007
http://www.omanet.om/english/geo/geog.asp, accessed 10 November 2007.

Mosley, John 2004
Starry Night Companion. Your Guide to Understanding the Night Sky. Toronto: Space Holding Corp.

MRM&WR 2008
Aflaj Oman in the World Heritage List. Muscat: MRM&WR.

MRMEWR 2002
Aflaj in the Sultanate of Oman. Muscat: MRMEWR

MRMEWR 2005
Water Resources in Oman. Muscat: MRMEWR.

Nash, Harriet 2007
"Stargazing in traditional water management: a case study in northern Oman", *Proceedings of the Seminar for Arabian Studies*, v. 37: 157-174.
URL: http://hdl.handle.net/10036/16852.

Nave, Rod 2007
http://hyperphysics.phy-astr.gsu.edu/hbase/eclip.html#c3
http://hyperphysics.phy-astr.gsu.edu/hbase/hframe.html (link)
accessed 30 July 2007.

Odeh, Mohammed 1998-2006
Arabic Star Names.
http://www.jas.org.jo/star.html,
accessed 14 June 2005
http://www.icoproject.org/star.html,
accessed 4 October 2008.

Orchard, Jocelyn 1995
"The origins of agricultural settlement in the al-Hajar region", *Iraq*, v. 57: 145-158.

Orchard, Jocelyn and Jeffery Orchard 2007
"The Third Millennium BC Oasis Settlements of Oman and the First Evidence of their Irrigation by *Aflaj* from Bahla", *Proceedings of the International Symposium Archaeology of the Arabian Peninsula through the Ages* 7th-9th May 2006. Muscat: Ministry of Heritage and Culture: 143-173.

Philip, George Limited 2000
Philip's Planisphere for Latitude 23.50' North. George Philip Limited.

Philip, George Limited 2003
Philip's Star Chart Epoch 2000. George Philip Limited.

Richards, Edward G. 1999
Mapping Time: the Calendar and its History. Oxford: Oxford University Press.

Schjellerup, Hans C.F.C. 1874
"Description des étoiles fixes composée au milieu du dixième siècle de notre ère par l'astronome persan Abd-al-Rahman al-Sûfi." *in* F. Sezgin, ed., *ᶜAbdarraḥmān as-Ṣūfī abu l-Ḥusain Ibn ᶜUmar (d. 376/986) Texts and Studies*. Frankfurt am Main: Institute for the History of Arabic-Islamic Science at the Johann Wolfgang Goether University: 5-282.

Selin, Helaine, ed. 2000
Astronomy Across Cultures: the History of Non-Western Astronomy. Dordrecht/Boston/London: Kluwer Academic Publishers.

Serjeant, Robert Bertram 1995a
"Ḥaḍramawt to Zanzibar: the pilot poem of the Nākhudhā Saᶜīd Bāṭayīᶜ of al-Ḥāmī", Paideuma, 28: 109-127", *in* R. B. Smith, ed., *Farmers and Fishermen in Arabia: Studies in Customary Law and Practice*. Aldershot: Ashgate Publishing Group.

Serjeant, Robert Bertram 1995b
"A Socotran star calendar" *in* R. B. Smith, ed., *Farmers and Fishermen in Arabia: Studies in Customary Law and Practice*. Aldershot: Ashgate Publishing Group.

Serjeant, Robert Bertram 1995c
"Star calendars and an almanac from south-west Arabia Anthropos, 5 49: 433-459", *in* R. B. ed. Smith, ed., *Farmers and Fishermen in Arabia: Studies in Customary Law and Practice*. Aldershot: Ashgate Publishing Group.

Al-Shaqsi, Saif R. 1996
Aflaj Management in the Sultanate of Oman. Case Study of Falaj al-Hamra (dawoodi) and Falaj al-Kasfah (aini). MSc thesis, University of Wales.

Al-Shaqsi, Saif R. 2004
The Socio-Economic and Cultural Aspects in the Implementation of Water Demand Management. PhD thesis, University of Nottingham.

Al-Siyabi, Hilal O. 2000
"Astronomy in traditional Omani agriculture", *PDO (Petroleum Development Oman) News*, v. 3: 20-27.

al-Sūfī, ᶜAbdul Raḥmān b ᶜUmar 995/1586
Kitāb Ṣuwar al-Kawākib al-Thābita (Illustrated Book of the Fixed Stars). Muscat: Ministry of National Heritage and Culture.

Al-Sulaimani, Zaher, Tariq Helmi, and Harriet Nash 2007
"The social importance and continuity of falaj use in northern Oman", *International History Seminar on Irrigation and Drainage,* 2-5 May 2007 Tehran. URL: http://hdl.handle.net/10036/15174

Tibbetts, Gerald R. 1971
Arabic Navigation in the Indian Ocean before the coming of the Portuguese, being a translation of Kitāb al-fawā'id fī uṣūl al-baḥr wa l-qawā'id *of Aḥmad b. Mājid al-Najdī*. London: The Royal Asiatic Society of Great Britain and Ireland.

Al-Tikriti, Walid Y. 2002
"The south-east Arabian origin of the falaj system", *Proceedings of the Seminar for Arabian Studies*, v. 32: 117-138.

Varisco, Daniel Martin 1990
"Folk astronomy and the seasons in the Arabian Gulf", *Al-Maᶜthurat al-Shaᶜbiyyah*, v. 19: 7 - 21.

Varisco, Daniel Martin 1994
Medieval Agriculture and Islamic Science. The Almanac of a Yemeni Sultan. Seattle and London: University of Washington Press.

Varisco, Daniel Martin 1997 a
Medieval folk astronomy and agriculture in Arabia and the Yemen. Aldershot/Brookfield: Ashgate Publishing.

Varisco, Daniel Martin 1997 b
"The Rain Periods in Pre-Islamic Arabia", *Medieval folk astronomy and agriculture in Arabia and the Yemen*, Section III. Aldershot/Brookfield: Ashgate Publishing.

Varisco, Daniel Martin 2000
"Islamic folk astronomy", *in* H. Selin, ed., *Astronomy Across Cultures: The History of Non-Western Astronomy*. Dordrecht/Boston/London: Kluwer Academic Publishers: 615-650.

Wehr, Hans 1966
A Dictionary of Modern Written Arabic, Wiesbaden: Otto Harrassowitz.

Wilkinson, John C. 1974
The organisation of the falaj irrigation system in Oman. Research Paper No 10, School of Geography, University of Oxford.

Wilkinson, John C. 1977
Water and Tribal Settlement in South-East Arabia. A study of the Aflāj of Oman. Oxford: Clarendon Press.

Wilkinson, John C. 1978
"Islamic water law with special reference to oasis settlement", *Journal of Arid Environments*, v. 1: 87-96.

Wilkinson, John C. 1983
"The origins of the aflaj of Oman", *Journal of Oman Studies*, v. 6: 177-94.

www.ingramcontent.com/pod-product-compliance
Lightning Source LLC
LaVergne TN
LVHW071219210926
848482LV00015B/850